CRITICAL GEOGRAPHIES

Edited by **Tracey Skelton**, *Lecturer in Geography, Loughborough University*, and **Gill Valentine**, *Professor of Geography, The University of Sheffield*

This series offers cutting-edge research organized into three themes of concepts, scale and transformations. It is aimed at upper-level undergraduates, research students and academics and will facilitate interdisciplinary engagement between geography and other social sciences. It provides a forum for the innovative and vibrant debates which span the broad spectrum of this discipline.

CULTURE/PLACE/HEALTH

*Wilbert M. Gesler
and Robin A. Kearns*

London and New York

First published 2002
by Routledge
2 Park Square, Milton Park, Abingdon, Oxon, OX14 4RN

Simultaneously published in the USA and Canada
by Routledge
270 Madison Ave, New York NY 10016

Routledge is an imprint of the Taylor & Francis Group

Transferred to Digital Printing 2005

© 2002 Wilbert M. Gesler and Robin A. Kearns

Typeset in Perpetua by
Florence Production Ltd, Stoodleigh, Devon

British Library Cataloguing in Publication Data
A catalogue record for this book is available from the British Library

Library of Congress Cataloging in Publication Data
Gesler, Wilbert M., 1941–
Culture/place/health/Wilbert M. Gesler and Robin A. Kearns.
p. cm. – (Critical geographies)
Includes bibliographical references and index.
1. Medical geography. 2. Social medicine.
3. Medical care – social aspects. 4. Epidemiology.
I. Kearns, Robin A., 1959–II. Title. III. Series.
RA792.G486 2001
614.4'2–dc21   2001034973

ISBN 0–415–19065–7 (hbk)
ISBN 0–415–19066–5 (pbk)

CONTENTS

CONTENTS

FIGURES

BOXES

ACKNOWLEDGEMENTS

It is a truism and yet worth restating that no book can be prepared in an intellectual vacuum. We therefore thank the various co-authors, faculty and graduate students with whom we have explored and developed some of the ideas surveyed in this book. For example, Chapter 8 draws on Robin Kearns's collaborative work with Ross Barnett. Over the past decade, we each have published papers in *Social Science and Medicine* and *Health and Place* that have subsequently advanced our own, and each other's, thinking. We therefore thank the respective editors, Bob Earickson, and Graham Moon and Michael Hayes, for their support over the years. Preparation of a book also draws on the goodwill and resources of host departments. We thank colleagues and staff in our respective Departments of Geography. Finally, we are fortunate to be part of a most convivial international network of health geographers, many of whom maintain an interest in our work. We are most grateful for this support.

The author and publishers would also like to thank the following for granting permission to reproduce images in this work:

David Carr for his photograph of a homeless person in Washington, DC (Figure 4.1).

Saatchi & Saatchi NZ/TV2, Auckland, and Communicado (producers) for the advertisement of the *Middlemore* television programme (Figure 8.4).

Rebecca Dobbs for her photographs of an access elevator and of the step up from platform to train (Figures 6.1 and 6.2).

Directions magazine of the New Zealand Automobile Association, for the photograph of roadside crosses (Figure 7.3).

Elsevier Science for Figure 8.1, in the article 'Consumerist Ideology and the symbolic landscapes of private medicine', from *Health and Place*, 1997, 3: 171–80.

Lucy Gunning for her photographs of the Roman Baths, Bath and the Royal Crescent, Bath (Figures 7.1 and 7.2).

Every effort has been made to contact copyright holders for their permission to reprint material in this book. The publishers would be grateful to hear from any copyright holder who is not here acknowledged and will undertake to rectify any errors or omissions in future editions of this book.

1

INTRODUCTION

Introduction

This book explores the links between culture, place and health from our vantage point as both cultural and health geographers. These three domains of social-scientific concern have independently, and in paired combinations, been of considerable interest to anthropologists, sociologists and geographers for some time. Recently, however, the so-called 'cultural turn' in medical geography has contributed to the transformation of this field into a new formulation as 'health geography'. The net result is an increasing interest in the way cultural beliefs and practices structure the sites of health experience and health care provision. Such sites are, from the geographer's viewpoint, best regarded as *places*, given the rich nuances of that term which direct attention to both identity and location (Eyles 1985).

Our efforts to draw together a focus on this tripartite set of themes is an extension of our earlier concerns with the processes of 'putting health into place' (Kearns and Gesler 1998). In that volume, we and our contributing essayists focused on difference in health outcomes and service opportunities, an established emphasis in medical geography dating from the challenges of political economy (Jones and Moon 1987; Eyles and Woods 1982). However, throughout the 1990s, emerging concerns centred on views of difference that went beyond issues of material well-being. We and colleagues were concerned with difference in places of health care as well as in the health experiences of groups defined by such markers as class, 'race', sexuality, or gender (or combinations of these identities) and how these differences might be made visible in research. The present book steps back from the earlier case study approach to provide a survey of key themes in the cultural geography of health and health care. Our goal is a book aimed at a senior undergraduate and beginning graduate audience that will provide an accessible survey containing comprehensive reference to current literature. We introduce our book by first considering the three domains we

1

seek to link. The discussion introduces many of the key themes we pursue in detail later such as structure and agency, narratives, difference, therapeutic landscapes, and consumerism. The chapter concludes with a glimpse at the changes taking place in health geography and a brief overview of the chapters that follow.

Culture

In his provocative volume, Don Mitchell (2000) asserts that while a decade ago, Peter Jackson (1989) offered *an* agenda for cultural geography, culture is now *the* agenda in human geography. Much the same can be said for health geography. Medical geography's experience of the so-called 'cultural turn' has been its reformulation and reorientation towards health (Kearns and Moon 2001). However, the precise agenda of health geography is far from clear, and cultural geographies of health are literally and figuratively 'all over the place'. Our goal is to more closely link cultural treatments of health in geography to social, political and economic forces. Drawing on Mitchell (2000: 6), we see that in the fields of health and health care there are 'arguments over real spaces, over landscapes, over the social relations that define places in which we and others live'. These arguments signal the fact that health is a contested term, and health care a contested terrain.

One does not need to look far for examples to substantiate this view that there is a cultural politics of health and health care. An enumeration of health concepts can produce a list of some twenty definitions ranging from the medical/easy-to-measure ('absence of disease') through to the 'new age'/difficult-to-measure ('wholeness'). In between, we can encounter variants that speak to economic as well as legalistic overtones ('a purchasable product', 'a right') (Lee 1982). The prevailing definition of health in any given place will shape what groups hold power over the processes involved. Contest over the 'ownership' of health leads to a power politics being mapped into the landscape of place through symbol, language or materiality. The fact that midwives cannot act independently in most hospitals, for instance, illustrates that the contest over women's health results in health care facilities in turn becoming a contested terrain. As another example of the cultural politics of health, Figure 1.1 speaks to polarized views on the moral geography of AIDS. With a single line of spray-paint, a second practitioner of graffiti has called into question whether sympathy should be shown to AIDS patients. We say this speaks to a 'moral geography' because we believe that place matters. The graffiti in Figure 1.1 was photographed in the very mixed and impoverished downtown East Side neighbourhood of Vancouver, and we can speculate that such contested views would be less likely to appear within established gay areas such as Cawthra Park near St Michael's Hospital and Casey House Hospice (see Chiotti and Joseph 1995) in Toronto. There, instead of outspoken

Figure 1.1 Graffiti, downtown East Side, Vancouver, 1994. Photo by Robin Kearns.

graffiti we encounter a more respectful and enduring memorial to Canadians who have died of AIDS (Figure 1.2).

Talk of materiality and cultural politics might suggest we are attributing too much to bodies and the 'bricks and mortar' of health and health care. On the contrary, we agree with Mitchell (2000) that there is no such 'thing' as culture, but rather that culture is expressed through 'the power-laden character and emplaced nature of social relations' (Berg and Kearns 1997: 1). Such assertions signal our affiliation with the so-called 'new' cultural geography. While by its very prefix, this term is implicitly contrasted to an 'older' tradition, we strive for an inclusive view of work in cultural geography that contests the new/old binary construct. One reason for this view is that, especially beyond North America, many cultural geographers have always had a political edge to their work (Berg and Kearns 1997).

We are interested not only in how culture affects health, but how health affects culture. Health, illness and medicine have reflected, and contributed to, the fashioning of modern culture (Lawrence 1995). Bury (1998) identifies some key processes that are carrying their influence into a postmodern era. One of these processes is *objectification*, in which private activities or knowledge are made more public, in the sense of becoming increasingly open, revealed and accessible. An example of this is sexual health, in which advertising for drugs such as Viagra has publicized and medicalized a condition formerly discussed only behind the

Figure 1.2 AIDS memorial, Cawthra Park, Toronto. Photo by Robin Kearns.

closed doors of a counselling room. For us, the relevance of this shift is that advertising literally takes place, adding to the landscape through billboards or contributing to everyday geographies through a range of media. A second process is *rationalization*, which Bury (1998) describes as the way in which modern medicalized life undermines self-identity through moving from the 'citadel' of the hospital or clinic into the community, thereby blurring the boundary between lay person and professional expert. For us the culture–place link is crucial in this observation. Efforts to calculate rationally the risks or the healthfulness of situations literally bring health home, and into other non-traditional healing places, thus fragmenting and re-placing voices of scientific authority. Through rationalization, patients have been (willingly or otherwise) transformed under recent expressions of capitalism into 'consumers': active purchasers of goods and services who have (often carefully circumscribed) rights of participation in policy debates. Our goal is to work, through an analysis of the linguistic and symbolic nature of 'culture', towards creating new understandings of 'the consumer', rather than simply adopting, adapting and testing explanations developed elsewhere.

Place

As geographers, we have as our foundational interest the idea of place. Most fundamentally, place is simply a portion of geographic space. Places can be thought of as 'bounded settings in which social relations and identity are constituted' (Duncan 2000: 582). These can be official geographical entities such as municipalities, or informally organized sites such as 'home' or 'neighbourhood'. The idea of place, and the associated terms 'sense of place' and 'placelessness', were strategically deployed by humanistic geographers in the 1970s to distinguish their work and intent from 'positivist' geographers whose work dealt primarily with space.

Recent scholarship has moved away from the uncritical 'humanistic' phase of the 1970s and towards formulations that consider the ways in which place is forged through intersections of local and global factors, as well as individual agency and societal structures (Jackson and Penrose 1993). Doreen Massey (1997), for instance, strives for a progressive and global sense of place that moves on from earlier preoccupations with the (lost) 'authenticity' of rural and pre-industrial places. The concept holds dangers, however. David Harvey (1989) suggests that interest in place can lead to a preoccupation with image through recourse to romantic myths of community. Similarly, from a gender perspective, Gillian Rose (1993) notes that the idea of (especially home) places as stable and secure can elide inherent inequalities.

Notwithstanding these concerns, we believe place is a useful organizing construct for developing a cultural geography of health. In a particularly useful formulation, John Eyles (1985) explores the interrelations between place, identity and material life. He describes sense of place as being constituted by two related experiences: that of actual, literal places, and that of 'place-in-the-world'. With the latter, he refers to the self or externally ascribed status that comes from association with, or occupation of, particular sites. Thus the urban resident may gain a sense of place through both the cumulative experience of urban space (such as his or her own neighbourhood) and the contingent feelings of esteem (or otherwise) that flow from that experience. This view is complemented by a later conceptualization by Entrikin (1991), for whom place is both a context for action and a source of identity, thus poised between objective and subjective realities.

Our specific interest is in the way cultural practices can be observed, and landscapes can be read as 'text'. We follow Duncan and Duncan (1988: 117) in drawing on Barthes's attempt to transcend landscape description and to instead 'show how meanings are always buried beneath layers of ideological sediment' Place has not universally been viewed in this way within health geography. Kearns and Moon (2001) point out that in Britain, for instance, it has been more of an implicit construct, with a distinction made between context and composition. There has been a tendency to reduce place to space and equate it with the ecological or aggregate (Moon 1990). Perhaps its most obvious manifestation – the study of therapeutic landscapes (places where place itself works as a vector of well-being) – has generated considerable international interest since Gesler's 1992

paper, with, most recently, the publication of an edited collection with contributors from four countries (Williams 1999). Beyond the work of geographers,
the idea of 'social capital' has become an implicit expression of the importance
of place in the role of supporting and promoting health (e.g. Wilkinson 1996).
The population health framework, for instance, illuminates the central role that
social geographies of everyday life such as housing, employment and social
networks play in shaping health status (Hayes 1999).

What of our place as authors? We write from academic bases in the United
States and New Zealand. As Kearns and Moon (2001) indicate, scholarly practice from beyond the non-English-speaking world surely has a place in speaking
to us. We must avoid creating a homogeneous view of 'an Atlanto-Antipodean
white health geography orthodoxy'. Yet in supporting this call, we (perhaps ironically) are two white males writing from the aforementioned double-A axis within
geographical scholarship. In doing so, we at least advocate bringing into the
project those from elsewhere in the academic world. Where possible, therefore,
we discuss the work of lesser-known writers and the perspectives of groups
whose voice is arguably less heard within the discipline. This may, of course, be
only a partial answer to bringing new people and places into the conversation
and the need to 'be audience to them, rather than implicitly expecting them to
be audience to us' (Kearns and Moon 2001: 7).

Health

The adjective 'medical' which has conventionally prefixed geographical scholarship concerned with health, disease and therapy has to some extent marginalized
our work into a domain perceived as perhaps too niche-like, applied and biosci-
entific. Health arguably allows an expansiveness that blends the cultural, environmental and social with greater ease. What do we mean by health? In our earlier
discussion of 'culture', we began the process of considering the range of 'cultures
of health' that exist. Another useful departure point is the word's origins in the
old English *haelth*. This derivation involved three meanings: whole, wellness and
the greeting 'hello' (Lee 1982). The connections between these meanings remain
with us today. In Western culture, a glass is raised at dinner and the toast is
'good health' Other cultures have more intimate forms of greeting. In Maori
protocol, for instance, the traditional greeting is the *hongi*, in which one's breath,
a necessity of life, is momentarily symbolically shared through the pressing of
noses.

The World Health Organization (WHO) (1946) signals this wholeness in its
often-cited definition of health as 'a state of complete physical, mental and social
well-being and not merely the absence of disease or injury'. This depiction of
health has been prefigured in indigenous belief systems such as those of First

Nations peoples in Canada (Stephenson *et al.* 1995) and Maori in New Zealand. For Maori, *hauora* (health) can be seen as a four-sided concept: the spiritual (*taha wairua*), the psychological (*taha tinana*), the social (*taha whanau*) and the physical (*taha tinana*) (Durie 1994).

Health is an intrinsically holistic concept and social phenomenon and one which easily becomes one of the 'metaphors we live by' (Lakoff and Johnson 1980), with broad application in situations beyond the body and biology (e.g. 'healthy cities'). Health involves more than personal wellness. It can be a metaphor that can 'steer place-making activities' (Geores 1998: 52). In this sense, healing places such as spas and thermal resorts can be promoted as such to the point that healing = place. Here, healing is associated with a bounded area in which therapeutic properties are to be found *here but not there*. Locations become conceptualized as containers that represent focal centres for healthiness and reputations found in, but not out of, place (Gesler 1992).

As Evans *et al.* (1994) point out, a vast amount of effort is centred on trying either to maintain and improve health, or to adapt to its decline. A subset of these activities, they say, involves the deployment of economic resources to the production and distribution of health care, that collection of goods and services believed to have a special relationship to health. Our concern is to explore a selection of those aspects of health and health care that contribute to, or are influenced by, place and local cultures. Our attempt to maintain a thematic breadth is consonant with the recognition of the reactive nature of the so-called 'health' care system. While we are interested in the cultural and place-specific aspect of sickness care, we recognize that the World Health Organization (WHO) rejected the absence of sickness as a *de facto* definition of health over forty years ago. The classic WHO definition of health arguably opens space for the cultural geographer to widen horizons to accommodate the blurred boundaries between medical concerns and the lay pursuit of well-being in society.

While 'progress' as an ideology is associated with the modernist mindset, the original WHO definition has been interpreted as 'unwittingly' postmodern in its depiction of health (Kelly *et al.* 1993). Paradoxically, the WHO, and agencies and individuals taking up its cause, have invariably opted for technical expertise in the quest for causes of ill health. Yet the positive view of health embedded in the WHO definition begs for an alternative view in which nuance, difference and contingency – hallmarks of postmodernism – surface onto the research agenda.

At the level of the individual, health as progress towards the development of a person's potential is a worthy goal. While we acknowledge the challenges of measurement, we believe it is better to adhere to a positive and challenging definition than to a contradictory yet measurable one. Health, we believe, must be considered a 'presence to be promoted and not merely an absence to be regretted' (Meade and Earickson 2000: 2). An implicitly postmodern perspective on health, anticipated by such commentators as Dubos (1959), celebrates the intangibility of health and, rather than despairing of its inherent relativism, acknowledges its roots in the moral values of individuals and political process. Our book continues a process of 'making space for difference' (Kearns 1996) in health geography by embracing an inclusive and expansive view of health.

As citizens and geographers we are both participants in, and observers of, turbulent times. In the health care sector, we witness people disillusioned by the commercial reorientation which sees patients recast as customers and a general striving to reclaim health as a quality less commodified, less medicalized and more connected to everyday life experiences (see Chapter 8). Such concerns, we believe, underlie a reawakening of interest in the notion of therapeutic landscapes (see Chapter 7).

Health is, to a large extent, constructed by the health care systems prevailing at any particular place and time. While the examples we use inevitably are drawn from within health systems, we are not especially concerned with the systems *per se*. Other recent surveys by geographers aimed at a similar teaching level (e.g. Curtis and Taket 1996) offer a comprehensive assessment of change within major and contrasting health systems such as those in the USA and Britain. In our book, the system is the stage on which relations between culture, place and health are examined, rather than the primary object of interest.

Narrating change

A key influence informing our book is the recent escalation of interest in narrative theory across the humanities and social sciences. This interest suggests a renewed willingness within Western scholarship to 'trust the tale' (Kearns 1997c). The narrative turn in health geography has manifested itself in a number of ways. First, there has been a (re)legitimation of the authorial perspective, manifested in the use of the personal pronoun. While perhaps apparently minor in the larger scheme of things, this does symbolize a return to 'geographer as teller of tales'. This willingness to be candid in reporting research reflects a renewed frankness about the way things are, and the experience of being researchers. Thus the place of safety (Dyck and Kearns 1995) and emotion (Widdowfield 2000) are now part of reflecting on the research process. We all know that personal disposition interacts with pragmatic considerations such as availability of funding or supervision, ultimately influencing choice of topics and approaches. However, as if story has been presumed to be best subordinated to science, there has been scant recognition of such matters. It is thus reassuring to note the personal becoming political at the micro scale of health research (e.g. Parr 1998).

An important development over the past decade has been that the 'twin streams' of medical geography (disease ecology and health services research) (Mayer 1982) have increasingly merged and, with newer strands of influence, become more like a braided river. Our book is intended to continue and consolidate the emergence of contemporary health geographies. In his 1993 paper, Kearns nudged the collective conversation towards a critical, if cultural/humanistic, standpoint through advocacy of what, at the time, was intended as a rather tongue-in-cheek term: 'post-medical geographies of health'. His call was an extension of a trend, noted at the beginning of the decade, towards: 'less concern about definitions of tight boundaries for the subject' (Bentham, *et al.* 1991). Following Kearns and Moon (2001), our argument is that although programmatic statements (e.g. Kearns 1993) may have a Kuhnian paradigm-shifting ring about them, in essence (social) scientific research programmes seldom shift rapidly or radically. Rather, it is more a case of evolution rather than revolution and, in developing links between culture, place and health, our book seeks to move the middle ground of dominant discourses within medical/health geography towards a firmer engagement with cultural concerns.

The remainder of the book is organized as follows. In Chapter 2, we define culture, then follow with a brief history of cultural geography to provide the reader with the principal concepts in the subdiscipline. In the past two decades, cultural geography opened out dramatically as it became informed by humanist, structuralist and postmodern theories and practices, so these ideas are discussed. Chapter 2 continues with a discussion of cultures of health and concludes by linking this idea with geography. Chapter 3 then steps back from the conceptual and surveys how we might research culture and health within particular places, granting special attention to how we place ourselves in the researcher's role. Chapter 4 digs deeper into the traditions of cultural geography, surveying the ways that humanistic and structural understandings converge in the cultural materialist perspective. Studies of unhealthy and deprived people are used as illustrations. Chapter 5 shifts focus to the linguistic ways in which culture and place are linked through health beliefs and behaviours. Words, we claim, are fundamental to the constitution of place and health care interactions, with metaphor being especially important in the way that one idea can stand in for another. This chapter includes an examination of research on naming, narratives, and the role of the media in health.

Recognition of difference, we argued earlier, has been a hallmark of the emergence of health geography over the past decade. In Chapter 6 we survey the place of cultural difference and the difference that place makes to different cultures. To make the ideas of this chapter concrete, we use illustrations from the literature on women's health, ethnic disparities in health care, gay and lesbian health experiences, and how people with disabilities (re)construct their lives. Chapter 7 is concerned with landscape and the way in which health and healing have been implicated in ideas of human–environment relations, social constructions of landscape, and personal landscapes of the mind. Ideas about nature as healer, marketing healing places, the role of symbols in healing landscapes, and therapeutic landscapes form the content of this chapter. We conclude with a chapter examining one of the dominant cultural constructions of Western capitalist societies: the health care consumer. By way of example, we consider the extent to which patients can in fact behave in a consumerist manner and the ways in which some health care businesses are attempting to lure patients to their places of provision.

2

CULTURE MATTERS TO HEALTH

Introduction

Along with Chapter 1, this chapter is about beginnings, about establishing some basic concepts that will be built upon throughout the rest of the book. Specifically, our task here is to look at two of the three concepts in our book's title, culture and health, and link them together. We begin by defining culture, with some cautions about using this rather nebulous idea. Then, in order to provide context to the practice of cultural geography today, we briefly trace the history of the subdiscipline from ancient to modern times. In two sections we discuss both an 'old' or traditional cultural geography and a 'new' cultural geography, the transition from one to the other, and a partial reconciliation of the two. The new cultural geography is informed by social theory, which leads into a discussion of three 'isms': structuralism, humanism and postmodernism. Our look at structuralism includes underlying forces that create divisions and inequalities in society along racial, ethnic, gender and other lines; the role of capitalism in creating and maintaining inequalities; and the socially relevant work of structuralist geographers. The humanist approach is characterized by a focus on subjectivity, experience and meaning, informed by the so-called 'philosophies of meaning'; practitioners employ qualitative, interpretive and ethnographic methods to 'get inside the heads' of the people they study. Our discussion of postmodernism examines notions of pluralism, difference and multiple voices; anti-narratives and deconstruction; and interpretive analyses of discourses. In the later part of the chapter, we turn to cultures of health, first discussing some specific ways in which cultural beliefs and practices affect the health of individuals and societies. Then we reinforce the culture–health link by showing how different aspects of health can be examined using some concepts from the old cultural geography, leaving approaches used in the new cultural geography for later chapters.

What is culture?

Culture is a notoriously difficult concept to define; we all seem to know what it is, and yet it is extremely difficult to put a definition down on paper. We can start by saying that culture includes the technology and organization that people employ to provide food, shelter and clothing within various physical and social environments. Using a taxonomy devised by Julian Huxley (Haggett 1975), we can say that it consists of material objects (artefacts) such as pottery, CDs and CAT scanners; social relations (sociofacts) such as kinship networks, racism and doctor–patient interactions; and ideas (mentifacts) such as incest taboos, democracy, and socialized medicine. Here is another definition:

> [Culture] is the complex of *socially produced* values, rules, beliefs, liter-
> atures, arts, media, penal codes, laws, political ideas and other such
> diversions by which a society, or any social group, represents its view
> of the world as its members (or at least the members in charge) believe
> it is or ought to be.
>
> (Lemert 1997a: 21; emphasis added)

Culture can be thought of as 'an inherited "lens" through which the individual perceives and understands the world that he [*sic*] inhabits, and learns to live within it' (Helman 1994: 3). This lens, for example, divides people into 'healthy' and 'ill', according to cultural norms. Anthropologists and others (including geographers) have emphasized the symbolic nature of culture. Thus Clifford Geertz (1973: 362) states that culture is 'a traffic in significant symbols' and Denis Cosgrove and Peter Jackson (1987: 99) define culture as 'the medium through which people transform the mundane phenomena of the material world into a world of significant symbols to which they give meaning and attach value'.

One must be very cautious when dealing with the concept of culture. First, it is not a thing; that is, it should not be reified as having a life of its own apart from the people who possess it. Second, culture must always be viewed in its historical, economic, social, political and geographic contexts. That is to say, the culture of a particular group of people at any one place and time is influenced by many other factors. A person's health status, for example, is often blamed on his or her 'culture', rather than on factors related to that person's social or economic situations (Helman 1994). Third, although culture is passed from generation to generation and may stagnate (giving what is termed *cultural inertia*), the more common situation is for culture to be constantly changing. This is an additional factor that makes it difficult to pin down what culture is. For example, a group of people who have beliefs about disease causation and treatment that include supernatural forces may come into contact with biomedicine. Fourth, we must acknowledge the possibility that, as Mitchell (1995: 102) claims, 'There's no such thing as culture.' Rather, he says, we should focus on how the *idea* (or ideology) of culture is used to control production and reproduction. Mitchell (2000: xv) shows 'how "culture" is never any *thing*, but is rather a struggled-over set of social relations, relations shot through with structures of power, structures of dominance and subordination'. Furthermore, culture should not be used as a source of explanation; rather, it is something to be explained as it is continually being socially produced by people as they struggle to achieve power and meaning.

A fifth caution in dealing with the culture concept, one that applies to geographers in particular, is that culture is possessed by groups of people of any size. Since having culture implies sharing ideas and practices, it is hard to think of culture as being confined to individuals, but a small group of friends or members of a household could be said to share a culture. Members of churches and social clubs certainly share a culture. Within most countries there are 'ethnic' groups with distinctive cultures such as West Indians in Britain or Vietnamese in the USA. Thinking in even grander terms, we can talk about French or Indonesian cultures or even a worldwide culture dominated by Western products and tastes. This latter popular culture is dominated by a capitalist economy that affects health beliefs and practices (Kearns and Barnett 1992, 1997). As examples, children's attachment to Barbie Dolls has been linked to bulimia, and recent studies show that although immigrant children coming to the USA have better overall levels of health than American children, their health soon deteriorates as their diets become 'McDonaldized'. At the same time as we recognize that culture exists among groups of many different sizes, we must also acknowledge that within any one geographic area multiple cultures or a *plurality* of cultures are manifest. There may be a few isolated tribes still in existence that possess a distinctive culture, but the vast majority of places contain people who are influenced by a myriad of cultural beliefs and practices. Finally, it should be recognized that definitions of culture, like cultural practices, are constantly evolving as the social contexts in which they are being made change. This is true also of definitions of what constitutes cultural geography.

A brief history of traditional cultural geography

The subdiscipline of geography that obviously deals with cultural matters and therefore includes cultures of health is cultural geography. To understand the role that cultural geography plays in the study of health today, it is important to understand its historical and social contexts. Cultural geography is a very old subdiscipline; we trace its history briefly here in order to appreciate its roots and changing emphases. That is, the practice of cultural geography today needs to be seen in the light of its own historical and social contexts to be fully appreciated. We tend to become so caught up in what is 'new' in geographic thought that we ignore the origins and evolution of key ideas. We know that people took an interest in the interplay between culture and the physical environment at least as far back as the classical Greek and Roman civilizations (Glacken 1967). One example is the Greek traveller and writer Pausanius, who, travelling in the second century AD, described in very accurate detail the beliefs and practices of the people who lived in different 'culture regions' of Greece. As Europe began to explore the world and came to dominate a large portion of it from the fifteenth century onwards, ideas developed that were to influence cultural geography. Giambattista Vico (1688–1744), for example, described a series of evolutionary steps in human culture, and Johann Herder (1744–1803) constructed his notion of the *Volk*, a collection of people who shared a common culture (James 1972).

The nineteenth century witnessed the origins of modern geography, cultural geography in particular. The two leading figures, Alexander von Humboldt (1769–1859) and Carl Ritter (1779–1859), were both concerned with human interactions with the physical environment. Among his other interests, Humboldt examined the relationship between population and natural resources, the origin of religions, and the dangers of deforestation. Ritter took a regional approach and tried to understand how human–environment interactions worked themselves out in different parts of the world (James 1972).

Humboldt, Ritter and others had a generally balanced view of the relative roles of humans and the physical environment; that is, they believed that neither dominated the other. However, as the nineteenth century came to a close, there was a conceptual shift toward *environmental determinism*, or the idea that the physical environment dominated or dictated human culture. One explanation put forward for this change in thinking is that as geography was becoming institutionalized as an academic discipline in universities, it was controlled largely by physical geographers who wanted to make sure that geographic study was 'scientific' and rigorous, and so they emphasized the importance of the physical environment and ways in which it could be measured quantitatively. There can be no doubt that the physical environment both constrains and provides opportunities for human activities; however, environmental determinists made untenable statements such as 'People who live in hot climates are lazy.' These kinds of ideas were used to rationalize European domination and exploitation of 'inferior' people who lived in non-Western environments (Livingston 1992).

One idea that was influential in early twentieth-century geography was *chorology*, *areal differentiation* or *region*. We can trace the origin of chorology to Ferdinand von Richthofen (1833–1905) and Alfred Hettner (1858–1941); the latter stated that

The goal of the chorological point of view is to know the character of regions and places through comprehension of the existence together and interrelations among the different realms of reality and their varied manifestations, and to comprehend the earth surface as a whole in its actual arrangement in continents, larger and smaller regions, and places.

(James 1972: 226–227)

The concept was used to develop *culture regions* in cultural geography, areas that were based on one or more culture traits such as language, religion, or type of agricultural practice. As time went on, many cultural geographers recognized that the culture region idea, although useful in putting some geographic order into the study of human cultures, nevertheless suffered from the difficulties that any attempt at regionalization encounters: decisions about categorization criteria, boundary definition, overlapping regions, and attempting to deal with more than one criterion.

Although environmental determinists such as Ellsworth Huntingdon (1876–1943) and Ellen Churchill Semple (1863–1932) did not believe that nature was entirely in control of human culture, their more extreme ideas caused geography a considerable amount of embarrassment. The pendulum began to swing in the opposite direction, towards a belief that human agency played a much more important role. In some cases, the extreme opposite view, *cultural determinism*, was taken; that is, that people with their technology and organization could dominate the physical environment. Most geographers, however, took a middle view, *possibilism*, which allowed for mutual environmental and cultural influences and recognized that the relative influence of one or the other varied a great deal over time and space. Paul Vidal de la Blache (1845–1918) is credited with originating the possibilist view. He believed that nature and culture were inseparable and developed his notion that *genres de vie* or ways of living both determined and were determined by various physical environments (James 1972).

In the 1920s, environmental determinism and chorology still dominated cultural geography, but in this decade a major turn was taken by the subdiscipline with the emergence of Carl Sauer (1889–1975), the 'father' of cultural geography in the USA, and the so-called Berkeley School. Although an American geographer, Sauer drew heavily on European geographers such as Vidal as well as American anthropologists such as Alfred Kroeber (1876–1960) and Robert Lowie (1883–1957). Sauer's great idea was the *cultural landscape*, a construct that included both the prominent features of the physical environment and transformations of the landscape through human activities. The cultural landscape constantly changed over time as various groups of people made their sequential impact. Sauer and his followers developed the landscape school of cultural

15

geography, one of whose dominant features was *cultural ecology*, or the processes by which humans alter and are altered by physical environments.

By the 1960s, cultural geography was a well-established subdiscipline within geography; in 1962 Philip Wagner and Marvin Mikesell in a landmark article summarized its content at that time. Defining cultural geography simply as 'the application of the idea of culture to geographic problems' (Wagner and Mikesell 1962: 1), they set out and discussed in detail what they thought were its major themes: (1) culture, (2) culture areas, (3) cultural landscapes, (4) cultural history, and (5) cultural ecology. Note that so far we have discussed all these themes except cultural history, which includes the idea of cultural evolution and the attempt to reconstruct the geographies of previous cultures.

In the 1960s and 1970s, positivist approaches (Box 2.1) and the so-called *quantitative revolution* came to dominate much of the field of geography. During this time, cultural geography tended to take a back seat to other subdisciplines, although a loyal band of scholars continued to produce very useful studies. At the same time, cracks were beginning to appear in the grand structure that Wagner and Mikesell had established in 1962. This is natural, and indeed desirable, in a field of study that wishes to remain current and dynamic. Harold Brookfield (1964) criticized cultural geographers for ignoring what he called the 'inner workings' of

---

Box 2.1 Positivist approaches

In this chapter we will be highlighting some philosophical approaches that geographers have taken over the past several decades. Each perspective will be described in terms of distinguishing features. Although positivist approaches have rarely been used by cultural geographers and are not emphasized in this book, they dominate much of what is done in geography today, including medical geography and some aspects of health geography. Therefore, geographers should be aware of their main elements.

Positivists use empirical methods in their research. They believe that knowledge is gained through direct experience of the world. The methods used by positivists are usually, but not always, quantitative; that is, they collect data on variables (e.g. physicians per 100,000 population) that are measurable. Positivists attempt to develop theories that are capable of verification, so they formulate hypotheses (e.g. physician-to-population ratios will be higher in urban as opposed to rural areas) that can be formally tested (e.g. using statistical analyses). Positivist geographers examine data to discern spatial structure and use methods of spatial analysis (e.g. location/allocation models to aid in the planning of health facilities) (Johnston 1986b; Gregory 2000a).

Starting in the 1970s, positivism was criticized on several grounds. Some scholars questioned the assumption that the supposedly objective methods used in the natural sciences were appropriate in the social sciences and humanities. Claims that science could be neutral or isolated from social life or that scientific knowledge was 'value free' were also challenged. Many felt that positivist science ignored both the contexts in which knowledge was acquired and the feelings and actions of people being investigated (Johnston *et al.* 2000).

geography, and said that much more emphasis should be placed on cultural processes. There were complaints that cultural geographers were paying too much attention to rural as opposed to urban areas, to material culture as opposed to ideas, and to folk rather than contemporary cultures. Even Wagner (1975) renounced his former statement made with Mikesell, saying that the five themes were too neatly packaged to cover the entire field of study adequately.

Geographers critical of traditional cultural geography took issue with scholars such as Wilbur Zelinsky (e.g. in his book *Cultural Geography of the United States*) who assumed that culture was a real force that 'existed "above" and independent of human will and intention' (Mitchell 2000: 30). The 'superorganic' status given to culture meant that it potentially had great power (e.g. to rally a nation around a perceived homogeneous culture). However, Duncan (1980) argued that cultural geographers had mistakenly reified culture as a thing, rather than a process, and ignored the role of individuals in shaping their own identities and destinies. Furthermore, Duncan and others found the old cultural geography to be irrelevant to modern life.

The emergence of a new cultural geography

Changes in the late 1970s and 1980s within geography (which paralleled changes in other social sciences) led to another major turning point in cultural geography. These changes were inspired by at least two developments: (1) the conscious incorporation of social theories such as structuralism, humanism and postmodernism; and (2) the development in Britain of social geography. Leaders of the 'revolution' in cultural geography such as Denis Cosgrove and Peter Jackson began to talk about a 'new' cultural geography that was opposed to the 'old'. Here is Cosgrove and Jackson's description of what was new:

If we were to define this 'new' cultural geography it would be contemporary as well as historical (but always contextual and theoretically informed); social as well as spatial (but not confined exclusively to narrowly-defined landscape issues); urban as well as rural; and interested in the contingent nature of culture, in dominant ideologies and in forms of resistance to them. It would, moreover, assert the centrality of culture in human affairs.

(Cosgrove and Jackson 1987: 95)

At the same time as the US tradition in cultural geography fathered by Sauer was being condemned as irrelevant, a group of young geographers, many of them from Britain, began to take an interest in cultural issues. A traditional cultural geography that focused on static customs and cultural artefacts was found to be inadequate for looking at the constant changes in cultural practices that occurred in the modern world. Mitchell (2000) traces the lineage of this thinking as the new cultural geographers attempted to deal with the 'culture wars' of the late twentieth century. These strategies included such events or issues as global and local economic restructuring, an increase in activist movements related to such problems as identity formation or land reform, and environmental degradation created by wasteful corporate practices.

A major source of ideas for the new cultural geography came from the emerging field of cultural studies that had its origins in the political, economic and social upheavals of post-Second World War Britain (Mitchell 2000). A key figure was Raymond Williams. Raised in a working-class family in Wales, Williams (1993) attempted to show that culture arises out of the ordinary institutions of everyday life. In his work he developed the concept of cultural materialism, elaborated by Peter Jackson in his book *Maps of Meaning* (1989). The view of cultural materialism was that everyday culture arose out of both the material or economic 'base' of life and such 'superstructure' elements as ideologies and social relationships. Another influential figure was Jamaican-born Stuart Hall. One of Hall's main interests was how one's identity was the product of contested relations of power. Here he drew upon the work of the Italian political theorist Antonio Gramsci (1971) and Gramsci's ideas about cultural hegemony. Gramsci tried to show how people who are subordinate in society come to consent to being dominated by others, even when it does not appear to be in their best interest. Dominant groups, he showed, develop ideologies that appear to make the order they impose seem natural. Subordinates, however, may resist the established order, and this results in struggle. The new cultural geographers were also influenced by postmodern ideas, described later in this chapter. A summary of key features of the new cultural geography can be found in Box 2.2.

Calls for a turn in cultural geography soon met with reactions from cultural geographers who had established reputations for producing solid and useful work. Lester Rowntree (1988) said, essentially, that it would be foolish to throw the

Box 2.2 Elements of a new cultural geography

It is difficult to evaluate just what constitutes the new cultural geography because, like culture itself, it evolves over time and may be defined differently by its various practitioners. Here we attempt to draw out some of the (overlapping) strands that can be found in current research. Many of these ideas are picked up later in this text.

1   An emphasis on process as opposed to form. As a reaction to the 'what', 'where', and 'when' questions addressed in the 'barn-type' traditional cultural geography, there is an attempt to answer 'how' and 'why' questions about the processes of culture formation.
2   An examination of how culture is constituted as part of everyday life. This includes examining closely forms of popular culture such as music, television, festivals and sports events. Here we find an interesting ambivalence or tension on the part of cultural geographers, who both sympathize with popular culture because it expresses the creativity and tastes of ordinary people and at the same time decry its commodification and commercial exploitation.
3   Framing culture formation in terms of politics. Some new cultural geographers would contend that what they are dealing with is cultural politics, the power struggles that dominant and subordinate groups engage in to define who they are and what their ways of life will be. There is talk of 'culture wars' (a phrase used by the conservative US politician Pat Buchanan) over such issues as abortion, gay rights, pornography on TV, and many other cultural issues.
4   A concern with identity formation. Culture wars are often about people's struggles to define who they are, in terms of, for example, gender (feminist politics), ethnicity (think of the Hutu and Tutsi in Rwanda) or nationality (think of the aftermath of the break-up of the former Yugoslavia). An important aspect of identity formation is how others dictate what should be done to one's body.
5   An engagement with concepts arising out of postmodern thought. These include listening to the plural 'voices' of various individuals and groups, exclusive and inclusive geographies, and rejection of meta-narratives.
6   A reaffirmation of the roles of space and place. New cultural geographers are excited about the ways in which cultural processes work themselves out over space and in particular places. They find that other

social science disciplines as well as the humanities share the same recognition that basic geographic concepts help to explain culture formation and change.

baby out with the bathwater; that is, he found much to admire in the new, but also wanted to preserve the old, and he stressed continuity rather than 'paradigm trashing'. Rowntree, Kenneth Foote and Mona Domosh (1989) echoed these sentiments, saying that the old/new dichotomy was false and that cultural geography was influenced by an 'epistemological pluralism'. In a sharp counter-attack, Marie Price and Martin Lewis (1993) accused the new cultural geographers of reinventing rather than revitalizing the subdiscipline, of stifling diversity, of attacking straw men and of misrepresenting the old. They wished to see a cultural geography that welcomed both social theory and human–environment relations. Within a few years, the dust of the debate seemed to have settled somewhat (as, perhaps, the dust is settling within medical or health geography today; see Chapter 1); an indication of this is a volume of readings (Foote *et al.* 1994) that blends new and old perspectives.

Our own approach in this book is to accept the value of a pluralism of cultural geographies; however, the emphasis is clearly on what is new. In the following two sections, we discuss three ideas from social theory that have informed the turn in cultural geography (as well as in human geography overall): structuralism, humanism and postmodernism. We realize that breaking down social theories into these three categories vastly oversimplifies what is an extremely complex area of thought. Most geographers who employ social theories in their teaching and research are influenced by more than one of these approaches. However, we make this division for the sake of clarity and exposition and because these three 'isms' have had the most direct influence on the cultural geography of health.

Structuralism and humanism

Although we agree with Rowntree (1988) that practitioners of the old cultural geography incorporated social theories such as humanism and structuralism in their work with little overt mention of theory, we believe that it is desirable to reveal one's theoretical underpinnings. Asking philosophical questions such as what one can know and how one can know it (epistemology) and what can be known (ontology) makes a great deal of difference in how we view the world and how we carry out research. For example, believing that research is valuable only if it lends itself to quantification is quite different from an emphasis on examining the value of subjective experience. No one is immune to making epistemological and ontological assumptions, despite the protestations of some scholars

20

that their work is neutral or completely objective. Some cultural geographers focus on one approach exclusively while others are eclectic, but we should be able to have some idea of 'where they are coming from'. Much of where the new cultural geography (and this book) comes from is social theory, by which we mean 'any theory of society or social life that distinguishes itself from scientific theories by its willingness to be critical as well as factual' (Lemert 1997a: 24). Two social theory approaches that have been used in the new cultural geography are structuralism and humanism, which focus on structure and agency, respectively.

Structuralist (often Marxist) thinking (Box 2.3), although engaged in by relatively few geographers, had an important impact on geography, starting in the early 1970s (Peet 1977). In large part a reaction to the dominant spatial science and positivist outlook of the time, it also was in tune with a period of political and social ferment in much of the Western world that began in the late 1960s. In the USA, people were beginning to protest against the conduct of the Vietnam war, and the civil rights movement was in full swing. Students and others in many western countries questioned established dominance hierarchies such as patriarchy and attacked the inequalities engendered by racism and capitalism. So-called 'radical' geographers chastised their colleagues for sitting in their academic ivory towers and shrinking away from attacking such problems as poverty, imperialism in foreign affairs, and women's rights (Bunge 1978).

Box 2.3 Structural approaches

Structuralism has its roots in linguistics and literary theory, political economy, anthropology and psychology (Gregory 2000b). The structuralist posits that there are hidden causal mechanisms that produce divisions and inequalities within society along the lines of class, ethnicity, gender, age and other characteristics (Johnston 1986b; Jackson and Smith 1984). Within groups of people, power relationships are established; some subgroups attempt to establish hegemony over others by persuading them (by force at times) to accept their dominant position and their cultural values. There is a very strong tendency for humans to create 'the other' and to elevate their own relative position by denigrating those who belong to another group; thus the European colonial powers spoke of the natives they conquered as being 'savages' and unfit for civilized society. Dominant groups (e.g. upper classes, whites, males, older persons) strive to legitimate their own positions (e.g. by passing laws favourable to themselves) and at the same time to marginalize others. A very important tool that is used to establish and maintain positions of power is developing *ideologies*

---

or ways of thinking that dominant groups attempt to establish as 'understood truths' (Jackson 1989). Examples would be the divine right of kings, the idea that 'A woman's place is in the home', and the notion that achieving wealth at the expense of others should be accepted as part of the 'American way of life'. It must be noted, however, that dominance is very often resisted by the subordinated and marginalized, resulting in conflict and struggle.

The idea that what appears on the surface in social relationships can be explained by hidden causal mechanisms can be traced back to nineteenth-century thinkers such as Auguste Comte (1798–1857), Emile Durkheim (1858–1917) and Karl Marx (1818–1883). Marx's structural approach focused on the idea of *modes of production* such as subsistence agriculture, which characterizes feudal Europe and much of the developing world today, or the factory system that developed as part of the industrial revolution that began in England and diffused around the world. It is a basic Marxist tenet that throughout most of the world today, modes of production are controlled by a capitalist economy. Capitalism creates a basic division in society, Marx said, between a dominant managerial class and a subordinate working class. Managers exploit cheap labour and alienate workers from the fruits of their labour (i.e. workers are not directly in control of what they produce). Furthermore, workers are dehumanized as their labour becomes yet another commodity to be bought and sold (Cloke *et al.* 1991).

Committed to socially relevant research, structuralist geographers attempted to reveal the mechanisms that led to hierarchies of power, dominance, ideologies and marginalization. Many became advocates for political, economic and social policies that attempted to reduce inequalities within society (see, for example, Bunge 1978). Although they were accused by their critics of being long on enthusiasm and short on hard data and impartiality, structuralist geographers carried out important work. As examples, studies showed how capitalism spatialized the economy by developing core areas of wealth and peripheral areas of poverty at international, national and local scales (DeSouza and Porter 1974; Harvey 1973); how the discipline of geography supported the exploitation of developing countries (Hudson 1977); and how social forces created urban ethnic segregation and severe inequalities in housing and the provision of services (Dear and Wolch 1987).

Structural geographers damaged their own cause by disagreeing among themselves about theory and methods, and many turned away from exclusively structuralist or Marxist approaches, but they did have a lasting impact on geography (including cultural geography). Even though many scholars felt that Marx's nineteenth-century ideas about the importance of class struggle were outmoded, they nonetheless held on to the idea that societal divisions and inequalities were important; as a result, geographers who investigated issues of gender, race and ethnic relations blossomed. Concepts such as hegemony and resistance, legit-imization and marginalization, and ideologies were used to help explain spatial patterns of human activities. There was a new sensitivity to the creation of 'the other', to how humans attempt to control space through territoriality, and to how even the idea of culture itself can be commodified and used to control production and reproduction (Mitchell 1995).

A second philosophical approach that emerged to contest the dominance of positivism was humanism (Box 2.4). Whereas structuralism was a new ap-proach for cultural geography, humanism had been a part of the subdiscipline from at least the nineteenth century. How do humanists carry out their work? A general approach used by many anthropologists and some geographers is called interpretive or *hermeneutic* research. Here one tries to interpret the experiences people have within the context of their daily lives. This requires trying to see the world from the point of view of others and so necessitates immersion into their activities and thoughts. The anthropologist Clifford Geertz (1973), one of the pioneers in the interpretive approach, advocates using the method of *thick description* or describing people's lives in as much detail and from as many angles as possible. There is a danger here, however, which is termed the *double hermeneutic*. That is, researchers are interpreting what a respondent told them, and what respondents say is, in turn, their own interpretation of events and ideas.

Box 2.4 Humanist approaches

A basic humanist tenet is that knowledge is obtained subjectively; human-ists attempt to understand personal experiences and feelings and how people attach meaning to their surroundings. In contrast to the structuralist focus on the overarching structures that constrain or provide opportunities for human action, the humanist examines human agency or the ability of indi-viduals to make decisions and change their environments. Humanist scholars recognize both the humanity of the people they study and their own humanity (Tuan 1976; Buttimer 1979).

The modern rise of humanistic thought can be traced to the Renaissance in Europe that began in the fourteenth century. As human geography got its nineteenth-century start as an academic discipline, it carried with it humanistic ideas. For example, Vidal's *genres de vie* had humanist content. However, in the 1950s and 1960s, as the quantitative revolution took

---

centre stage, humanistic ideas were pushed into the background; examining human feelings was not considered by many scholars to be a scientific enterprise. Then, in the late 1970s and 1980s, the pendulum began to swing back in favour of humanistic approaches as dissatisfaction grew with the inadequacy of positivist models to explain human experience. It was realized that the messiness and ambiguities of human life often could not be quantified; people's actions could not always be predicted by models based on 'rational' principles (Ley and Samuels 1978).

As humanism re-emerged as an approach used by cultural geographers and other groups, it had to develop theoretical stances and methodologies for research. Theory was derived from the so-called *philosophies of meaning*, including idealism, pragmatism, phenomenology and existentialism. *Idealists* hold that what we know ultimately derives from subjective experience, and thus in order to understand what others believe and think, we must somehow 'get into their heads', using the method of *verstehen* or 'understanding'. The *pragmatist* believes that people act on the basis of how they interpret the world; through a process of trial and error they decide what is best or has the most utility for them. *Phenomenologists* try to understand themselves by examining closely their personal experiences and consciousness, and speak of discovering the true essence of objects by stripping away both common-sense and scientific views of the world. Rather than essence, *existentialists* emphasize being or existence. They claim that humans are free to choose the nature of their existence and to give it meaning, but that people must also assume responsibility for what happens to them (Cloke *et al.* 1991; Johnston 1986b). Unfortunately, the language used by philosophers of meaning is often difficult to penetrate; however, the focus on human agency, meaning and personal experience is clear.

Humanistic research usually employs qualitative methods (for examples, see the 'Focus on qualitative approaches in health geography' in *The Professional Geographer* 1999, vol. 51 (2)). A standard approach is to conduct *in-depth interviews* with study subjects; the interviewer asks questions that are loosely guided by research questions, but the respondent is encouraged to talk freely to reveal his or her feelings. Interviews are often taped and transcribed and then a rigorous process of analysis is applied to pick out themes and patterns from the scripts. Some researchers actually live with their subjects for a period of time, engage in activities with them, and record the details of their interactions, a method called *participant observation*.

It is hard to imagine cultural geography without the humanistic approach; indeed, reviews in the journal *Progress in Human Geography* have been titled 'Cultural/humanistic geography'. Much fascinating and useful work has been done in which the humanistic strain is clearly discernible:

two examples will have to suffice. David Ley (1988) lived for several months in an inner-city African American neighbourhood in Philadelphia, gaining valuable insights into how residents thought and acted, often belying the perceptions of them by outsiders. Graham Rowles (1978), using a technique he called *experiential field work*, spent several months interviewing just five older persons to understand in a very profound way how they viewed the spaces within which they lived.

Both structuralism and humanism, as we have seen, had important impacts on the practice of cultural geography. Not surprisingly, advocates of one approach or the other did not always appreciate the others' point of view and arguments ensued, epitomized by the structure/agency debate. Which was most important, scholars asked, the underlying structures of society that dictated human actions or the human capacity to make decisions and alter structures? Part of the answer was, of course, that it depends on the situation. More and more, however, it was realized that both structure and agency must be taken into account in most situations and should be seen in a dialectical relationship to each other. Attempts were made to reconcile humanist and structuralist ideas in new approaches, notably *structuration* and *realism* (see also Lemert 1997b). We leave this issue for the moment, but will return to it in Chapter 4. Meanwhile, we turn to a very controversial topic, postmodernism.

Postmodernism

Postmodernism is a notoriously difficult term to define; it is many things to many people. Some scholars embrace it and others approach it with fear and loathing. Like the pluralities it celebrates, it can take on a plurality of meanings. To begin with, its very name indicates both continuity and change: it can be seen as both a continuation of the modern, another phase of the modern (which began, in the Western world, a few hundred years ago); or it can be seen as distinct, something in opposition to the modern. Some see the postmodern period as a time when what we thought of as the 'modern' world is experiencing a series of very serious crises of global proportions as the Euro-American colonial systems have collapsed, the gap between rich and poor widens, and new social movements have arisen based on female, ethnic and gay rights (Lemert 1997a). Here we will try to extract some themes of postmodernity about which there appears to be some consensus. Box 2.5 does the same for a related philosophical approach, poststructuralism.

Certainly, one of the most distinguishing features of postmodernism is its emphasis on plurality, recognizing that there are multiple 'voices' in society that

Box 2.5 Poststructuralism

Poststructuralism refers to the thinking of a group of French theorists including Jean Baudrillard, Gilles Deleuze, Jacques Derrida, Michel Foucault, Julia Kristeva, Jacques Lacan and Jean-François Lyotard (Pratt 2000a). Poststructuralists reacted against the claims to objectivity and truth of structuralism. Here we attempt to encapsulate some of their principal ideas, which will be seen to overlap with those of postmodernism (Best and Kellner 1991; Blunt and Wills 2000; Pratt 2000a).

1   An emphasis on language 'as the medium for defining and contesting social organization and subjectivity' (Pratt 2000a: 625). Language, it is argued, does not reflect reality; rather, it constitutes it.
2   A focus on the formation or identity of the subject. It is recognized that identities are historically produced. They are performed in the activities of one's daily life, rather than being naturally given. Also, the subject, rather than being a unified whole, is decentred; that is, attention is paid to the multiple, interacting facets of the subject such as gender, sexuality, class, ethnicity and degree of disability (see Chapter 6).
3   A focus on power and its intimate connection with the formation of knowledge. Power relations are seen as diffused throughout society, affecting social relationships between individuals and between individuals and society. There are multiple forms of power that control the actions of people and they are contested through the politics of feminism, sexuality, class and ethnicity.
4   A repudiation of thinking in terms of the binary oppositions of Western philosophy such as subject/object, male/female, appearance/reality, speech/writing, reason/nature, as such thinking puts the inferior term (e.g. female, nature) in a negative light. A good example of this is Lawrence Berg's study of the way dichotomous thinking leads to 'the social construction of a hegemonic masculinity to effect a specific geographic understanding of the world' (Berg 1994: 245).
5   An insistence that 'there are no facts, only interpretations, and no objective truths, only the constructs of various individuals or groups' (Best and Kellner 1991: 22). Attempts to impose certain ideas (e.g. monogamy or patriarchy) as natural are contested.

may lay claim to their own 'truths' and should be listened to. We can illustrate this idea by taking postmodernism to be both an object of study and an attitude (Box 2.6) towards the knowledge that can be gained about the world (Cloke *et al.* 1991). One of the best examples of postmodernism as *object* is found in architecture (see Figure 2.1 for an example of a postmodern hospital). Postmodern architecture is often a reaction to the rational, efficient, cold buildings designed by such modern giants as Le Corbusier (1887–1965) or Mies van der Rohe (1886–1969); it is characterized by a variety of styles (sometimes parodies of previous styles), giving rise to the expression *pastiche*.

Box 2.6 Postmodernism as attitude

Postmodernism as an *attitude* denies that any one theory can explain how the world works (Smart 1996). This flies in the face of the European 'Enlightenment project', which, through rational thought and scientific experiment, sought to understand how nature worked and humans behaved. However, throughout the modern period there have always been those who rejected grand schemes, theories about everything, especially when they didn't work. It is clear to many people that what the West promised to the rest of the world simply could not be achieved (Lemert 1997a). One of the consistent themes of postmodernism is an opposition to *meta-narratives*, a scepticism about single explanations of everything, an insistence on a plurality of explanations. Meta-narratives, it is argued, are often attempts by certain groups to dominate others by imposing their own interpretations of the world (note elements of both the dominance idea from structuralism and the interpretive idea from humanism here). Accepting the foregoing ideas, however, leads to a problem that has been termed the *crisis of representation*, which refers to the extreme difficulty a researcher has in adequately understanding or depicting other people and places (Clifford 1988).

The method used to counter meta-narratives and hegemonic interpretations is called *deconstruction*. Deconstructionists claim that there is no universal language that is a perfect reflection of reality; there are no unmediated truths about the world (Boyne 1990). Perhaps the most familiar example of deconstruction is in the field of literary criticism, where the idea is that a novel, poem or play can be interpreted in an almost infinite variety of ways. This idea, however, can be extended to any social, political, economic or cultural phenomenon. Thus one could offer multiple explanations for racism, the failure of communism in Eastern Europe, the economic crisis in South-East Asia, or the popularity of the film *Titanic*.

Figure 2.1 Interior of Starship children's hospital in Auckland, New Zealand. This example of postmodern architecture simulates a spaceship to playfully diffuse the stereo-typical austerity of hospital buildings and appeal to younger patients. Photo by Robin Kearns

Some would take the modern city as an example of postmodernism as object. Thus Edward Soja (1986: 270) describes Los Angeles as 'a confusing collage of signs which advertise what are often little more than nominal communities and outlandish representations of urban location'. Society itself can be analysed as postmodern object. Seeking to understand the economic, political and social chaos that is often noted in contemporary society, David Harvey in his book *The Condition of Postmodernity* (1989) argues that the disorder is due to 'the logic of late capitalism'. That is, capitalism has reacted to its recent problems by replacing the rigid structures and practices of the 'Fordist' period of production and consumption with a much more flexible strategy that employs modern communication technologies and credit mechanisms to enable capital and entrepreneurs to move rapidly around the world, attempting to keep a step ahead of new crises.

A prominent theme in postmodern thought is the importance of *discourses*, language or texts; that is, close attention is paid to the ways in which a society expresses thoughts and ideas. Michel Foucault (1965), for example, examines in detail how madness was thought about by Europeans in different historical time periods. Here again we encounter the problem of knowing what a particular discourse, text or use of language really means, as what is said or written can be interpreted in many different ways. A key idea along these lines for geographers has been the notion that landscapes can be 'read' in the same way as literary texts (Duncan and Duncan 1988). The Duncans argue that human landscapes represent the transformation of ideologies into concrete form. As geographers, we attempt to read landscapes, but we must realize that 'Interpretations are the product of social contexts of historically and culturally specific discourses; they are constructed by interpretive communities and they frequently, but not always, reflect hegemonic value systems' (p. 120). We take up these ideas again in Chapter 5.

What are the negatives and positives of postmodernism for cultural geographers? On the minus side, critics of postmodernism, with some justification, claim that it is shallow, ephemeral and depthless. There is also the complaint by some that postmodernism fosters an 'anything goes' attitude, that one can interpret a cultural phenomenon in any way one chooses, without having any standards for judging quality. This tendency toward fragmentation, it is argued, leads to chaos and there is no way to judge whether research is good or not. However, few scholars are willing to abandon standards and many, if not most, would advocate taking positions on, say, specific social issues such as institutional racism or women's rights. Donna Haraway's (1991) concept of *situated knowledges* speaks to this issue of taking stances. David Harvey, in his book *Justice, Nature and The Geography of Difference* (1996), also makes a passionate plea for taking positions, in this case on environmental justice.

We find much that is potentially useful in postmodern ideas, particularly those that contribute to the themes of this text. Cultural geographers by the very nature of their subject have a tradition of examining different cultures and can sympathize with the postmodernist celebration of *difference*. There is an awareness among cultural geographers that it is necessary to examine a plurality of cultures, and to be constantly aware of the problem of ethnocentrism or viewing other ways of living only through the lenses of one's own culture. In other words, cultural geographers must pay heed to other narratives when investigating such topics as environmental perception or the causes and treatment of diseases.

Of utmost importance to the authors of this text is the renewed emphasis that postmodernism puts on *place*. Abandoning the attempt to explain the world through meta-narratives or what is happening everywhere, it speaks of understanding local narratives and what is going on in places. Cultural geographers have always written well about what happens at micro scales, so in a way this is nothing new. However, we contend that now there is much more to say about places, given the new cultural geography. In Chapter 6 we focus on both difference and place.

Cultures of health

What if we set up a biomedical health care system, complete with high-tech diagnostic equipment and excellent geographic accessibility to the population it aimed to serve and nobody came? It is doubtful that this would ever really happen, but the question highlights the possibility that the biomedical health culture and the culture of the population to be served might be so incompatible that people would not use the service. In other words, culture matters: what people believe and practise in the realm of disease and its treatment is very important to the overall health of a population. The connection between culture and health is what we wish to explore here.

First, a definition. 'Health culture', Weidman says (1977: 25),

> refers to all the phenomena associated with the maintenance of well-being and problems of sickness with which people cope in traditional ways within their own social networks. It is a general term that includes both the cognitive and social aspects of [health traditions]. The cognitive dimension involves values and beliefs, the blueprints for health action, and requires us to understand theories of health maintenance, disease etiology, prevention, diagnosis, treatment and cure. The social system dimension refers to the organization of health care or the health care delivery system. It requires understanding of the structure and functioning of an organized set of health-related social roles and behaviors.

This definition covers the medical anthropologist's and the medical sociologist's concerns. The traditional, positivist health geographer would add that health culture includes the spatial arrangement of various aspects of belief systems (e.g. sites where health beliefs arise) and components of health care delivery systems (e.g. cultural influences on the locations of practitioners). More recently, health geographers would emphasize such things as people's experiences of illness and treatment in specific places, how health cultures vary from region to region, and how culture interacts with the environment to affect one's health.

In what specific ways does culture have an impact on health? For a start, ideas about such medical matters as anatomy and physiology arise from culture and affect health (Helman 1994). Examples abound: an emphasis within a society on slimness may lead to anorexia; balancing the bodily humours has been a persistent theme in many medical systems (e.g. Greek medicine); the Qollahuaya Indians of the Bolivian Andes liken their body's systems to the hydraulic cycles of the mountain on which they live and thus fear practices such as taking blood (Bastien 1985); rural North Carolinians may use the lubricant WD40 to 'penetrate to the joints' as part of their treatment for arthritis (Arcury et al. 1999);

and Foucault (1965) describes how cultural definitions of who is mad and what should be done about madness have changed over the centuries.

Studying the ways in which culture affects health is not easy because it involves people's thoughts, feelings, experiences and values. Social scientists of health, however, have devised some concepts and methodologies for making inquiries into the cultures of health. A basic concept is *explanatory models* (EMs) of health (Kleinman 1994) (Box 2.7). EMs are hard to pin down because they are changeable and idiosyncratic. They are sometimes consciously brought forward to deal with a problem and sometimes simply 'understood truths' that are outside a person's awareness. Everyone engaged in a medical episode (the patient, family and friends, practitioners of various kinds) has an EM; this helps to explain why patients with their lay knowledge of disease and biomedical personnel with their focus on scientific logic often clash in medical encounters, thinking and talking past each other. EMs should always be examined within their social, economic, political and cultural contexts. For example, asking a person why they think they got HIV/AIDS might involve their educational level, who within their social networks they obtain information from, and their religious beliefs. Whether or not a person seeks medical care for a work injury might depend on their income, the availability of health insurance, and their degree of pride in their 'toughness'.

Box 2.7 Explanatory models

Explanatory models (EMs) are used to explain (1) the aetiology or cause of a condition, (2) the timing and onset of symptoms, (3) the pathophysiological processes involved, (4) the natural history and severity of the illness, and (5) the appropriate treatments for the condition (Helman 1994). Each individual employs his or her own EM to deal with particular illnesses; however, many specific ideas are also shared by members of the same culture.

The example of a typical EM relating to arthritis will illustrate the characteristics set out above. It is believed by many people that arthritis is caused by exposure to cold and wet conditions, perhaps through one's work or other activities, although there is little scientific evidence that this is so. Few people are able to express the pathophysiological process involved; they only know that their joints become very stiff and sore. Some apply a mechanical model and talk about their joints needing to be lubricated, as one would a machine. Most people have difficulty tracing the history of their arthritis, but they may be acutely aware of its severity, often complaining that the pain is worse under wet and cold conditions. Finally, many treatments are tried, including, as would be expected, attempts to warm up one's joints by rubbing them or applying heat.

The method that is commonly employed to study health cultures is *ethnomedicine* or ethnomedical science (Fabrega 1975, 1977). Fabrega (1975: 969) states that ethnomedicine is 'the study of how members of different cultures think about disease and organize themselves toward medical treatment and the social organization of treatment itself'. Although ethnomedicine has often been used to study 'other' medical systems, it can also be applied to biomedicine, the dominant system in many parts of the world. Ethnomedicine has been used to study a wide range of health behaviours. For example, it is thought by some scholars that when sickness is attributed to sorcery this represents a social sanction on unacceptable behaviour rather than bad feelings among individuals (Rubel and Hass 1995). Others have shown that sickness and healing can be used for purposes of social control. Some ethnomedical studies have focused on how stories or narratives about illness reveal EMs (Price 1987). Investigators have also looked at what types of personalities healers possess and how healers are recruited and trained (Helman 1994).

Cultural geographies of health

To bring the geographic perspective into the discussion, we revisit the themes of cultural geography discussed previously and apply them to health cultures. Although this book focuses mainly on ideas from the new cultural geography, we conclude this chapter by providing illustrations of how some concepts from the old cultural geography can be used in the study of health, and pick up on newer ideas such as structuralism, humanism and postmodernism in subsequent chapters. Although we find the newer ideas to be more relevant to health care situations, there are also valuable ideas to be derived from cultural geographies that have yielded useful results in the past (see Gesler 1991).

Health cultures often vary from one area to another. Health geographers and others have used the idea of *culture regions* to think about such topics as the distribution of practitioners from different medical traditions in the states of India (which vary significantly in terms of language, religion and other culture traits) (Bhardwaj 1980); about local 'physician cultures' that lead to surprisingly large differences in surgery rates for the same diagnosis in small areas very close to each other (Wennberg and Gittelsohn 1982); and how Thai, Chinese and Western health cultures help to shape the areas of Bangkok, Thailand, in which there are different types of practitioners (Techatraisak and Gesler 1989). Studies like these have had to confront the problems of defining medical culture boundaries and overlap of medical culture areas. Still, the areal differentiation concept helps to bring some order into *medical pluralism*, the confusing array of alternative health cultures that are often available to health care consumers.

Turning now to *cultural ecology* or human–environment relationships, we find a strong tradition of disease ecology in the work of medical geographers who focus on spatial patterns of disease, but not as strong a tradition in health care delivery studies. Both environmental determinist and cultural determinist positions have been taken by health cultures from at least the time of Hippocrates. As just one example, the Hippocratic writings blame the sterility of the Scythians on both their cold climate and their riding of horses (Glacken 1967). A major concern of cultural ecology has been adaptation to changing physical and social environments. One can ask, for example, how societies throughout the world are adapting to the AIDS epidemic. In the UK and the USA, health care systems are adapting to changes in political and economic environments by various schemes that involve more privatization and 'managed care' (Scarpaci 1989a). Even the locational decisions of physicians can be examined from an adaptation perspective (Rundall and McClain 1982).

The *cultural history* theme is clearly applicable to health. The history of biomedicine as it evolved over space and time involves, among other things, origins in various centres of learning in ancient Greece, Roman influences, inputs from Arab scholars in the Middle Ages, new scientific advances beginning in the European Renaissance, and discoveries of new medicines from plants found by European travellers. In the Western world over the past century the history of doctor–patient relationships showed a dramatic change from having a family doctor living just down the street to visiting the doctor in a medical complex in a hospital or shopping mall, influenced by several economic, social, political and cultural changes within society (Knox *et al.* 1983). The recent history of changes in health care delivery systems in Western countries and how these changes affect both urban and rural areas are fascinating (see, for example, Mohan and Woods 1985; McLafferty 1986; Gesler and Ricketts 1992).

Our final traditional theme is *cultural diffusion*. Health ideas and health products originate at one or more places and are dispersed over space and time, helped by physical and human carriers, hindered by physical and human barriers. Take the case of computed tomography (CT) scanners. A culture of health care commodification and quick profit-making, coupled with local hospital high-tech cultures led by radiologists, helped to move the new product very rapidly across the US medical landscape between 1973 and 1977, following a classic S-shaped diffusion curve (Baker 1979). A micro-scale cultural diffusion of a health idea can be seen at work as knowledge about birth control spread through social networks within South Korean villages (Rogers 1979). And both native and biomedical health cultures are clearly manifest in Bolivia as information about oral rehydration therapy (ORT) moved both hierarchically and contagiously through a chain that included the WHO, the Ministry of Health, clinics, physicians, pharmacists, traditional healers and market vendors to peasants (Weil and Weil 1988).

Culture matters!

A popular book that helped to reinvigorate geography in the 1980s was titled *Geography Matters!* (Massey and Allen 1984). We borrow from that idea to claim that culture matters, an idea that anthropologists live by, but which geographers need to be more aware of. More specifically, we make the point in this chapter that culture matters to health. What people believe and do about illness and its treatment are crucial to an understanding of health. This is only a beginning to our discussion of how culture matters to health. In subsequent chapters we will apply the approaches advocated by the new cultural geography to this theme. We will also show with numerous examples how culture matters to health in places.

The chapter began by attempting to define culture and, at the same time, urged caution in the use of the term. We suggested that readers avoid thinking about culture as a thing or a variable with which to explain other phenomena (e.g. health status) and focus instead on understanding how new cultures are continuously being produced. We presented a brief history of cultural geography in order to provide a context to the practice of that subfield today. Today one thinks of such concepts as culture regions, cultural ecology, cultural diffusion and cultural landscapes as traditional or 'old' cultural geography, but these concepts can still inform the geography of health, as the previous section illustrates. What interests us now, however, is concepts arising from the 'new' cultural geography.

What is probably of greatest importance to anyone who wants to carry out research in health geography is the discussion of philosophical approaches in this chapter: structuralism, humanism and postmodernism. Our treatment of these social theories was necessarily brief and readers are encouraged to follow up on ideas that interest them. The incorporation of social theories created an immensely exciting ferment in health geography; no one can seriously do health geography work today without a sound knowledge of the 'isms'. We often pick up on the ideas that inform current cultural geography in the following chapters. Chapter 3 surveys methods adopted within the qualitative turn in health geography, which are in large measure adapted from anthropology, the disciplinary home of culture. As Chapter 4 focuses on structure and agency, it naturally comes back to structuralism and humanism. Chapter 5, a discussion of language in health geography, derives much of its theoretical basis from postmodernism and poststructuralism. Discussions of various differences in Chapter 6 are also informed by concepts from the two 'posts'. The landscapes of healing idea (Chapter 7) is mainly built upon cultural ecology, structuralism and humanism. In Chapter 8, we survey the ways consumption has become a dominant orientation of Western culture, and indeed health care itself.

Further reading

Harvey, D. (1989) *The Condition of Postmodernity*, London: Blackwell. Harvey's book is a clear explanation of postmodern society. Geographers and non-geographers alike have been strongly influenced by the ideas developed in this classic.

Helman, C. G. (1994) *Culture, Health and Illness*, third edition, Oxford: Butterworth Heinemann. This text is an excellent, very readable introduction to how anthropologists look at disease and health issues. Especially useful are its many examples used to illustrate the main concepts.

Jones, K. and Moon, G. (1987) *Health, Disease and Society: An Introduction to Medical Geography*, London: Routledge and Kegan Paul. This text was the introduction for many medical/health geographers to the possibilities for a theoretically informed study of health and illness.

Ley, D. and Samuels, M. S. (1978) *Humanistic Geography*, Chicago: Maaroufa Press. This book served to introduce humanistic thought to many readers in the early days of the reinvigoration of this philosophic approach.

Mitchell, D. (2000) *Cultural Geography: A Critical Introduction*, Oxford: Blackwell. A mandatory read for those who would like to understand the new cultural geography and recent thinking about contested formations of a wide variety of cultures.

3

STUDYING CULTURE/
PLACING OURSELVES

Introduction

The types of research we have discussed in the preceding chapters have stressed new ways of seeing the links between culture, place and health. Here we use the word 'seeing' metaphorically to signal apprehending, comprehending and inter-preting. But literally seeing is important too, and, in fact, a critical art of the cultural geographer. To see is not just to recognize what is there, but to notice connections in the observable landscape or 'secondary' textual materials (e.g. advertising material). However, while visual observation is a key to many types of research, there is more to observation than simply seeing. It can also involve a more complete sensory experience: touching, smelling and hearing the environ-ment as well as implicit or explicit comparisons with previous experience (Rodaway 1994). Further, we consider it important to acknowledge that seeing implies a vantage point, a social and literal place in which we position ourselves to observe and be part of the world (Jackson 1993). What, and how, we observe from this place is influenced by whether we are regarded by others as an 'insider' (i.e. one who belongs), an 'outsider' (i.e. one who does not belong and is 'out of place'), or someone in between.

The goal of this chapter is to reflect on the reasons why observation and inter-pretation are central to research in the cultural geography of health. We also wish to consider critically what it means to observe the in-place cultures of health and health care. The chapter explores various positions the researcher can adopt vis-à-vis whatever is observed (whether this be a set of buildings, some advertising material, or people interacting in a clinic). Following Kearns (2000), we argue that observation and interpretation have been taken for granted as 'natural' processes which therefore do not need the attention granted to technical procedures under-taken by health geographers such as survey design or population sampling. We maintain that with the added ingredient of critical reflection, observation can be transformed into a self-conscious, effective and ethically sound practice.

We begin by considering how the new cultural geographies of health introduced in Chapter 1 reorient the researcher towards experience and direct
observation of the world. We then more closely consider the place of observation in research. Next we discuss participant observation, a research approach in
which the researcher frequently occupies a 'grey zone' between involvement and
detachment. This section deals with participant observation as involving the
researcher immersing herself or himself in the context of the place being studied.
We then consider the power dynamics of fieldwork as an embodied activity. A
final substantive section steps aside from reflecting on observing people to consider
briefly how landscape and associated texts might be read. Our concluding reflections lead us into the remaining chapters of our book, which survey substantive
themes linking culture, place and health.

Surveying the field

Medical geography has tended to be a 'magpie discipline', collecting what fits
for theory-building from elsewhere (Kearns and Moon 2001). The subtle shift
currently taking place is that practitioners are increasingly crafting their own
theory and reshaping our views of how the world works. Closely connected with
this 'theorizing' has been a gradual shift in the construction of 'the field' for
cultural geographers of health. Given that the roots of the tradition are tropical
disease ecology (Meade and Earickson 2000), then the field was (and for some,
still is) a place that is *there* rather than *here*. From this perspective, data is collected
in the field (there), whereas theory is studied, then applied to analysed data, in
the academy (here). As Duncan (1993: 42) remarks, the practice of fieldwork
came to be named and practised 'in order to professionalise it and thereby elevate
its products above the representations of amateurs'. A key shift among cultural
geographers of health has been a drawing on feminist scholarship, which contends
that we are 'always, already in the field' (Katz 1994). Thus recent cultural geographies of health have less frequently been accounts of the exotic, such as tropical
health systems (e.g. Good 1987) and more often critical interpretations of the
ordinary conditions of our own (largely urban, Western) experience (e.g.
Williams 1999). At the same time, we encourage health geographers (preferably
non-Westerners from perspectives within their own societies) to study culture,
health and place in non-Western settings.

One might argue that all research involves observation, or at least is made up
of a series of observations. Thus in a social setting, a researcher might metaphorically 'observe' the population by employing questionnaires through which he or
she establishes the frequency of certain variables (e.g. occupation, sex). The
activity of conducting a survey invariably places the researcher in the position of
an 'outsider', marked as 'other' by purpose, if not appearance and demeanour

(e.g. clothing, age, ethnicity, or type of language used). Identifying a sample and coding the responses to questions involves concerns related to the goal of generalizability to broader populations. But the cost of these concerns is that the subtleties of culture are missed. The very nature of a survey means that only a restrictive subset of social phenomena are deemed to be of interest, and even these are isolated from their context, and generally unaccompanied by less directly observable values, intentions and feelings. Cultural geographers have tended to opt for qualitative approaches such as interpreting the narrative accounts yielded by in-depth interviews (see Box 3.1). These allow the consideration of human experience, suspending traditional positivist concerns about researcher bias and recognizing instead the relationship between the researcher and the people and places he or she seeks to study.

Box 3.1 Qualitative methods for health geographers

The techniques used by traditional health geographers were mostly quantitative (e.g. calculating practitioner-to-population ratios for geographic areas, measuring distance-to-care, and location/allocation modelling). As health became more of a social science topic and health geography became more informed by social theory, there was a movement towards using qualitative methods. Techniques include (1) in-depth interviews with individuals and groups; (2) participant observation and other ethnographic methods that study a culture intensively; and (3) interpreting such 'texts' as landscapes, archival materials, maps, visual images, and literature (Johnston *et al.* 2000).

Recognizing the usefulness of qualitative methods in health geography, Susan Elliott and others organized sessions at the annual meeting of the Association of American Geographers in 1997, resulting in the publication of eight papers on qualitative approaches in the focus section of *The Professional Geographer* (May 1999, vol. 51(2)). In her introduction to the papers, Elliott (1999) made the important point that one's choice of research methodology should depend on the research question asked. More and more, health geographers are asking questions that such techniques as in-depth interviews and participant observation can be used to answer. The *Professional Geographer* papers are a good place to start when looking for examples of qualitative work in health geography. They include studies of physician–place integration, professional relationships between family physicians and hospital specialists, community development approaches to heart-health promotion, using storytelling to understand women's desire

---

to tan, and the utility of in-depth interviews for studying the meaning of environmental risk to individuals.

A discussion of qualitative methods is especially appropriate for this chapter because these techniques are often used to identify (a) structure in the form of the social, economic and political forces that impact on people's health; and (b) agency in the form of revealing people's experiences of illness and health care in places (Mohan 1998). To put this another way, qualitative methods are useful in answering questions about context and meaning. They tend to shift the focus from biomedical concerns to alternative ways of knowing (Dyck 1999). For example, a study of people with HIV/AIDS in a particular urban context can serve to shift attention from the spatial patterns of the disease and its ecology. The social forces that affect people with the disease (e.g. prejudice, high costs of drug treatments) and how individuals cope with HIV/AIDS (e.g. how they cope with stigma and disability) are placed in the foreground in such work. Anyone who has carried out qualitative research will realize that it involves a series of negotiations with study subjects (Wilton 1999). The preconceived notions of the researchers, the settings for interaction, and the relationships established between researched and researcher can affect the outcome of the research as well as the people involved. Researchers therefore need to think carefully about how they are going to collect qualitative data (see also Curtis *et al.* 2000).

The place of observation

There is a range of purposes for observation in health research. Two of these can be summarized by the words 'counting' and 'contextualizing' (Kearns 2000). The first purpose – counting – signals an enumerative function for observation. For example, we might accumulate observations of the number of clinics in a neighbourhood, or even the number of patrons arriving at a health-promoting site such as a gymnasium or spa in order to establish daily rhythms of activity within these places. Under this observational rationale, other elements of the immediate surroundings are intentionally ignored so as to focus on a particular activity. The resulting numerical data can then be given graphical representation or subjected to statistical analysis. This approach to observation may be useful for establishing trends, but is ultimately too reductionist to develop in-depth cultural understanding of place.

A more comprehensive purpose of observation might be called contextual understanding. The goal in this approach is to construct interpretations of a particular time and place through direct experience. To achieve such under-standing the researcher immerses herself or himself in the socio-temporal context of interest and uses first-hand observations as the prime source of data. In this situation, the observer is very much a participant.

We emphasize that these two purposes (counting and contextualizing) are not mutually exclusive. One can observe with mixed motives and seek, for instance, both to enumerate and to understand context during a period 'in the field'. In research at outpatient clinics in the Hokianga district of New Zealand, for instance, the second author both counted the occurrence of waiting room conversations, and assessed how the topics discussed related to the surrounding context (Kearns 1991). In using this example, we are highlighting a concern that is obvious, but occasionally overlooked: observation also includes listening. Effective listening can assist visual observation by both confirming the place of the researcher as a participant and attuning the researcher to the sonic geographies or aural aspects of social settings (Smith 1994).

Observation through participating/participating while observing

Developing a cultural geography of health and place requires us to move beyond reliance on formalized interactions such as interviews. For no matter how much we are able to put people at ease before, during and after an interview, its struc-tured format often removes the researcher from the 'flow' of everyday life. An interview usually has an anticipated length and occurs in a mutually agree-able place often set apart from other social dynamics. In contrast, less structured forms of interaction such as participant observation are concerned with devel-oping understanding through being part of the spontaneity of everyday interactions.

Participant observation is most closely associated with social anthropology (Sanjek 1990), but the approach has been adopted and adapted by geographers seeking to deepen understanding of the meanings of place and the links to health experience (e.g. Gesler 1996) (see Box 3.2). Early examples such as Mick Godkin's (1980) research with recovering alcoholics have been updated with work by mental health geographers (Pinfold 2000; Parr 2000). Although talking to 'locals' in the field is not unusual, the depth of their involvement, their recurrent level of contact with people, and a set of relatively unstructured interactions marks this work as different from that undertaken by other health geographers. The key issue is that these researchers are more interested in *experience* than in perception or behaviour.

Box 3.2 Finding a betweenness of place in the field

One of the challenges of such work is the 'presentation of self' (Goffman 1963) while in the role of researcher. Adopting a research role and presenting oneself in mental health care settings is a particularly vexing task, and Vanessa Pinfold's (2000) work in Nottingham is illustrative. She conceptualized herself in the field as a visitor within a medicalized landscape that defined mental illness, but when writing she cast herself *between* a service user and professional. Pinfold's need for such 'betweenness' (Katz 1994) was in part predicated by the prevailing culture of (mental) health care systems. If culture is taken to be constituted by power politics (Mitchell 2000), then the prevailing (sub)cultures of doctors and patients are sufficiently potent to necessitate being seen as (at least partially) legitimate to each. However, as Pinfold (2000: 202) relates in the following dialogue with a patient named Michael, an attempt at remaining detached yet legitimate may sometimes break down:

Michael:	What do you want? Come to see the nutters have you ha, ha, ha?
Vanessa:	Michael? Why do you say that?
Michael:	Mad, mad, mad you see.
Professional:	I don't think he's in the right frame of mind today to do this [the research interview], we'll try again next week.
Michael:	Nutter, that's what we all are, come to see us . . . go away.

The slippage of roles apparent in the above passage from Pinfold's research log can potentially be overcome with participant observation, which involves a more wholesale immersion into a research context. Few, however, have chosen mental health care settings for this approach, given the sort of challenges discussed by Pinfold (but see Estroff 1981). Writers tend to refer to examples of participant observation rather than to offer step-by-step guidelines because every ethnographic situation is unique. Peter Jackson defines participant observation as a 'conscious and systematic sharing, in so far as circumstances permit, in the life activities and, on occasion, in the interests of a group of persons' (1983: 39). Thus it is the intentional character of observations that contrasts the activities of a participant observer to those of routine participants in daily life (Spradley 1980). To generalize, participant observation for health geographers involves strategically placing oneself in situations in which systematic understandings of place are most likely to arise. By way of example, Gesler's (1996) research on Lourdes as a healing place was informed by a degree of participant observation, given his decision to travel with a group of pilgrims to that destination and participate, where appropriate, in their collective experience. His interpretation was thus quite different from that which might have occurred had he simply visited Lourdes as a tourist, for participation in the social processes being observed increases the potential for more 'natural' interactions to occur.

Embodied power, knowledge and observation

Being a complete participant is difficult to achieve, and perhaps not desirable, for a certain critical distance is needed to maintain a research viewpoint. Similarly, it is difficult to imagine how being a complete observer might be incorporated into geography. However, whatever one's positioning vis-à-vis 'the observed' it is important to acknowledge that the act of observation is imbued with power dynamics. In this section, we consider how these power dynamics arise and the role of our bodily presence in contributing to the cultural politics of place.

In ethnographic attempts to understand the culture/health nexus, the body has a new-found recognition as an 'unstable site' from which to interpret the human world (Parr 1998: 28). Parr (1998) points out that with only limited exceptions (e.g. Crang 1996; Longhurst 1995; Hall 2000), the physical body has been neglected in writing about methodology and the construction of geographic knowledge. Yet the body (and not just any body, but our bodies) do mediate research contexts, opening or closing avenues of investigations for us. On account of such bodily markers as age, sex and 'race', we may be deemed 'in' or 'out' of place (Cresswell 1996). In some situations, our bodily difference can prompt us to consider the cultures of the places in which we are conducting research (see Box 3.3 for an example). According to Parr (1998: 35), to become unclean, and even unhealthy, is sometimes necessary as part of research practice and, symbolically and physically, it forces us to confront that which, in certain respects, we might consider to be 'other'.

To observe the human world involves participating – socially and spatially. We cannot observe directly without being present, and our bodily presence brings with it our personal characteristics such as 'race', sex and age. Belonging to dominant groups in society can mean that we potentially carry with us the power dynamics generated by such affiliation. Being a white adult male, for instance, will invariably create challenges to being a participant observer in the midst of a group whose members do not share those characteristics, such as a new mothers' support group. In such instances, collaboration in a mixed research team with

---

Box 3.3 Dropping in, blending in

The foregoing observation resonates with the experience of the second author, who undertook research through a period of partial participant observation, volunteering at a drop-in centre for chronic psychiatric patients (Kearns 1987). The air there was blue with cigarette smoke and the settings precipitated feelings of unhealthiness. Yet being well known to centre users required a commitment of time spent on the terms of those present. Elsewhere, while the author was attempting covert participant observation at a rural health clinic in a predominantly Maori district, his attempts to 'normalize' a research presence by partial concealment behind a newspaper were broken by a *kuia* (female elder), who entered, and greeted all present with a kiss. With the (news)paper wall removed between researcher and researched, the bodily presence of 'the other' was fully revealed. A shift in the research dynamic occurred, and different observations then occurred through interaction rather than a covert gaze (Kearns 1997a).

---

gender-appropriate interviewers is clearly preferable. In other words, our difference in terms of key markers of societal power (or lack of it) contributes to our ability to be 'insiders' and participants in the quest to understand place.

More subtle challenges to research dynamics may be generated by our level of education and affiliation to academic institutions. Foucault (1977) recognizes that social control can occur through the use of space, and particularly the ways in which one group (the powerful) are able to maintain watch over members of another (the disempowered). Foucault's ideas have implications for cultural-geographic research on health issues. The 'surveillance' he refers to is a very visual and dis-embodied form of observation. Gillian Rose (1993) has linked surveillance to the traditional geographical practice activity of fieldwork. To her and other feminist geographers, geography has been an excessively observational discipline characterized by an implied 'masculine gaze'. A challenge, therefore, is to be observers who go beyond the visual in our apprehension of the world. The key point is that observation is a power-laden process deployed within the context of institutionalized practices. As researchers based in, and representative of, academic institutions, it is imperative that we be aware of the ways in which others' behaviour may be modified by our presence.

The challenge posed by feminist geographers and anthropologists is to see field-work itself as a gendered activity (Rose 1993; Nast 1994). Their argument is that in being participant observers, we unavoidably incorporate our gendered selves into the arena of observation. As Kivell (1995) points out, observation in human geography is far from a simple matter of unidirectional watching. Rather it involves interactions that are potentially laden with sexual energy, and the resulting situations may shed light on the gendered constitution of the field site itself (see also Parr 1998).

Such comments do, however, presuppose that one has access to a field site. Gaining access to a research situation will be straightforward if one has a known role, such as Crang (1996) did when he worked as a waiter. However, there are often no convenient roles in health care sites, so once institutional ethical approval is gained, perhaps there is good reason to resist being typecast into any role except that of outsider. This ambiguous position worked for the second author in the health clinics of the Hokianga district of New Zealand (Figure 3.1). For, unable to pass legitimately as either a health professional or a patient (for the membership of both groups were too well known), being a visitor reading the newspaper helped maintain a reasonably inconspicuous presence for a short period (Kearns 1991).

The impact a researcher has on those encountered will determine, to a large extent, the ease with which they incorporate him or her into their place. Embedded within the word 'incorporate' is *corpus*, the Latin word for body. This reference signals the idea of the researcher's embodiment, the recognition that in being researchers we take more than our intentions and notebooks into any situation: we also take our bodies (Longhurst 1997). The way we clothe ourselves,

Figure 3.1 Pawarenga clinic, Hokianga, New Zealand: a research setting requiring modified participant observation. Photo by Robin Kearns.

for instance, can be a key marker of who we are, or who we wish ourselves to be seen as, in 'the field'. By way of example, research on the experiences of psychiatric patients in an inner-city setting might mean that the choice of wearing older clothes to drop-in centres minimizes the chance of being regarded as yet another health professional or social worker intruding on their lives. However, a troubling question is whether such a 'blending in' could be attempted if one were studying more elite social relations of place (e.g. the dynamics of a medical practitioners' convention). This matter of the power dynamics within research relations is a further reminder of the necessity to keep issues of conduct to the forefront of our thinking. For now, it is worth noting that our ability to relate to others in the field depends not just on appearance, but on the level and type of activity one undertakes.

A further point is that although concern for appearance and clothing is appropriate, it potentially reinforces geographers' fixation with the visual which Rose (1993) has emphasized. Hester Parr (1998) describes how, while researching the geographical experience of people with mental illness in Nottingham, she became conscious of non-visual aspects of her 'otherness' such as smell. Subtle differences distinguishing her from the patients at the drop-in centre such as wearing perfumed deodorant served to set her apart and inhibit the chance to interact freely. Parr's reflection on her observational experiences serves to remind us that senses such as smell add to the character of place (Porteous 1985) and merit consideration for the way they can mark as 'other' the bodiliness of the researcher (see Rodaway 1994).

A related point influencing field relations is that codes of behaviour are attached to different settings. Parr (1998) describes how she established a rapport with Bob, an acutely schizophrenic man with whom drop-in centre workers agreed that communication was difficult. The setting for this 'breakthrough' was a city park, an example of neutral ground which neither the researcher nor the research subject regarded as home ground. This site, at which social boundaries were blurred, contrasted with the drop-in centre, where roles adhered to established routines. The lesson is that research relations may be enabled or constrained by the (often unspoken) ways in which social space is codified and regulated.

A recognition of the regulation of social space suggests two types of rules in undertaking observation: first, to act ethically and responsibly; and second, to be attentive to, and respect, the (often unwritten) cultural codes of conduct maintained by the people we are interacting with. It is generally agreed that cross-cultural fieldwork is more problematic. This is for two reasons highlighted by the metaphor 'field'. First, the researcher is potentially venturing onto another's turf. Second, fieldwork involves a researcher working 'in a field' of knowledge, so there is the risk of eliding local understandings and priorities, and of (possibly unintended) one-way traffic of knowledge from the field (periphery) to the academy (centre). In such situations, the development of 'culturally safe' research practice is important. This practice emphasizes recognizing the ways in which collective histories of power relations may impact upon individual research encounters. Culturally safe research also stresses the need for appropriate translation of materials into everyday language, and the return of information to the communities that generated it (Dyck and Kearns 1995).

Interpreting texts

Health care landscapes and their representations can also be a focus for observation. The symbols and 'iconography' of landscape (Cosgrove and Daniels 1988) can be 'read' for their underlying ideological significance. Symbols are central to understanding sense of place, which in turn can be understood as the feelings of connection people have with specific locations (see Chapter 8). Landscapes – both on the ground and in their representation – may be regarded as repositories of cultural meaning (Cosgrove and Daniels 1988). The iconographic method seeks to reveal alternative cultural meanings by describing the form and content of landscape. Through this approach, researchers recognize that '. . . landscape meanings are unstable over time and between different groups, always negotiated, and political in the broadest sense' (Cosgrove 2000: 366). This call to view landscapes in their spatio-temporal contexts can be seen in work on spas as therapeutic landscapes as reviewed in Chapter 7 (Gesler 1998). In another example, therapeutic landscape ideas have been applied in an analysis of the Starship children's hospital in Auckland, New Zealand. This work analyses aspects of its built form, drawing on the present observable form of the hospital as well as various briefing papers and drawings by architects and designers (Kearns and Barnett 1999).

Language is also a key constituent of landscape (see Chapter 7). In a recent paper, Gesler (1999) advocates the analysis of the language of medical encounters. We discuss these ideas of place-creation through language more fully in Chapter 5. One specific form of language is advertising, which can be *interpreted* as well as simply described or enumerated by the cultural geographer (Jackson and Taylor 1996). A depth of interpretation is possible because it is a specialized form of communication whose function is not only to inform, but to persuade (Jefkins 1994). Analysts of advertising suggest that interpretation should examine *signifiers* (the things portrayed) in order to illuminate the *signified* (the underlying ideas) (Williamson 1978). Leiss *et al.* (1986) summarize two major methodologies that have been employed in the study of advertising messages: semiology and content analysis. Semiology concerns itself principally with the relationships between parts of a message in the belief that 'it is through the interaction of component parts that meaning is formed' (Leiss *et al.* 1986: 150). Semiology is derived from the traditions of linguistics and literary criticism. In contrast, content analysis developed from social science approaches and involves identifying and enumerating the occurrence of chosen phenomena in written or pictorial texts. Content analyses of advertising, for instance, attend to issues of sample size and reliability that might be of little concern to the semiologist.

In summary, the links between culture, place and health can be studied through conducting 'readings' of the relationships within images and other textual material to reveal the construction of meaning in landscape. Data sources for this sort of exploration might include the collection of visual images projected into both the printed media and the built environment by promoters of health care services. These promoters might be agencies seeking to recruit custom, or groups seeking to register protest as funding priorities (see Figure 3.2, and Chapter 8 for further examples).

Conclusion

In this chapter we have surveyed a range of ways in which we might examine the links between culture, place and health. We have emphasized observation as a unifying theme for how we might practically approach research questions. Observation, as we noted earlier, is a key human sensory capacity and one that is (at least implicitly) harnessed in most forms of research. We have stressed relatively unstructured forms of observation that have the capacity to yield depth, if not breadth, of insight. However, in identifying counting alongside contextualizing, we acknowledge the complementarity of quantitative and qualitative approaches within research designs.

We concur with Elliott (1999) that in exploring the links between culture, place and health, the research question should determine the method. Yet some methods bring us into closer contact with 'the researched' than others. We therefore took time to deliberate on the power dynamics of research. For just as some field research into landform processes involves very real physical safety concerns, so too there are cultural safety issues when one is moving closely into contact with others for research purposes (Dyck and Kearns 1995).

We close by asking whether there are disadvantages to reliance on observation. One danger, perhaps, in relying on experience 'in the field' is privileging face-to-face community relations while less localized relations remain beyond view (Gupta and Ferguson, 1997). To believe only what we see would be to deny the existence of social structures and processes occurring 'out of sight' (e.g. health-promoting social support via long-distance relationships, or seeking medical advice via the Internet). A further concern that is often raised is 'How do we know that the fruits of participant observation are valid?' Evans (1988) reminds us any method is, to a degree, valid when the knowledge it constructs

47

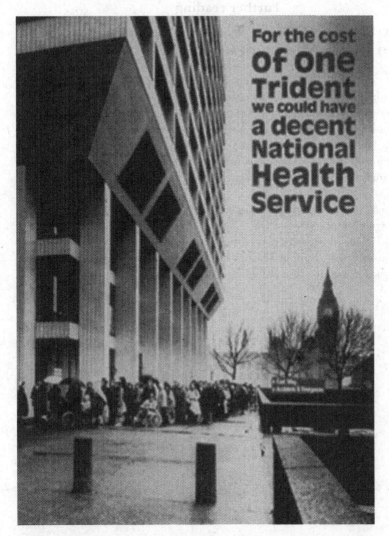

Figure 3.2 Health care as contested political terrain. In this British postcard, note that elements of the text can be 'read': the bold statement, the line of people waiting outside the hospital, the juxtaposition of the British Parliament build‐ing with the hospital in the foreground.

is considered by stakeholders to represent the social phenomena that it seeks to understand. Having considered some approaches to information-gathering, we survey in the following chapters some of the key traditions and contemporary research concerns at the interface of culture, place and health.

Further reading

Hay, I. (ed.) (2000). *Qualitative Research Methods in Human Geography*, Melbourne: Oxford University Press. This very readable book includes contributions on power and ethics, design and rigour, interviewing, focus groups, textual analysis and computer-assisted analysis – in short, a comprehensive range of topics useful to the cultural geographer dealing with health issues. The chapter by Kearns titled 'Research through observing and participating' elaborates on some of the themes raised in the foregoing discussion.

The Professional Geographer, Focus: Qualitative approaches in health geography, (1999) 51: 240–320. Box 3.1 mentioned the section of a recent issue of *The Professional Geographer*. This is a useful resource for both discussions of conceptual issues and the presentation of case studies on such topics as researching HIV/AIDS and tanning behaviour by women.

Baxter, J. and Eyles, J. (1997). 'Evaluating qualitative research in social geography: establishing "rigour" in interview analysis', *Transactions of the Institute of British Geographers*, 22: 505–525. A common criticism of qualitative approaches has been that they lack 'rigour' compared to 'hard' quantitative studies. In this influential article, Baxter and Eyles argue for ways to assess and maintain rigour in qualitative work.

4

STRUCTURE AND AGENCY

Introduction

Think for a moment about your last encounter with the biomedical health care system, say a visit to your family physician or general practitioner. What factors were involved in your experience? Here are just a few of the things you might consider: whether your visit was on a private or public scheme, how you paid for the visit, and whether your social status made a difference in the care you received. You might also recall how comfortable you felt in the doctor's office, whether or not you agreed with her about what was wrong with you, and whether you were frightened or reassured by the medical technology you encountered. The first set of factors can be thought of as the result of medical system-wide political, economic and social *structures*; the second set the result of the meanings produced by individual *humanistic* concerns. We first encountered the roles of humanistic and structural factors in culture in Chapter 2 and left it at that. Now we take them up again and see how they apply more specifically to cultures of health. The structural perspective is illustrated with a section on unhealthy societies and the humanistic perspective with a consideration of the situation of deprived people. We will also show how both types of approaches have been used together in *structuration* studies. The structuration approach can be aided by some ideas that come from time geographies of health behaviour, the next topic treated in this chapter. Finally, a blend between structure and agency is illustrated with a section on mental health.

The structural approach

There is no one structural approach; however, there are several key concepts that typify these approaches (Box 2.3). The structural perspective assumes that there are underlying forces in society that create divisions along the lines of ethnicity or race, class, gender, age and other population characteristics. Political, economic, social and cultural institutions reflect these divisions. Social forces lead to dominance/subservience relationships, creation of 'the other', legitimizing one's own group and marginalizing outsiders, and resistance to domination. Applied to health care, this approach examines, among other topics, (1) inequalities in service provision; (2) the medicalization of care; (3) the effects of restructuring and privatization; and (4) resistance to dominance.

Perhaps the most obvious manifestations of social forces at work in health care are the various types of political and economic systems in place(s) around the world, which go a long way towards determining the quantity and quality of care available to different groups within each country. The *welfare state system* in which health care is administered by a (usually) democratic government is perhaps best illustrated by Sweden, but Britain's National Health Service and medical care in other Western countries also are heavily involved in this approach. China and Cuba are both good examples of *socialist systems* where, again, there is very strong centralized control over health resources by socialist governments. In a *free enterprise system*, there is competition by various groups to supply health services; the USA is the exemplar here, although Britain, Canada, Australia, New Zealand and many others incorporate free enterprise elements. *Developing country systems* are a mixture of systems, including free enterprise, welfare, socialist, and strong components of traditional professional (e.g. Chinese, Ayurvedic) and traditional non-professional (e.g. herbal) medical systems. (See Curtis and Taket, 1996, for further discussion of health systems.)

Within countries or sections of countries, a great deal of evidence shows that different sectors of society do not receive the same amounts or quality of care. Although the structuralist approach was not invoked as a theoretical basis for their work, medical geographers in the 1960s and 1970s carried out studies that demonstrated these *inequalities* in geographic terms (as examples, see Bashshur *et al.* 1971; Morrill *et al.* 1970). More recently, there has been explicit recognition by health geographers of the underlying social forces that create inequalities, often expressed in terms of the impact of the capitalist economic system on health care provision. Along these lines, Jones and Moon (1987) discuss the ability of dominant groups to obtain better health care than others by legitimizing existing capitalist structures and reproducing capital accumulation processes, and several scholars have looked at the effects of capitalism on health in developing countries (Box 4.1).

Another aspect of the structural approach that has received a great deal of attention in recent years is the *medicalization* of health care. The dominant biomedical paradigm has been very successful in imposing its explanatory models (EMs) of disease and disease treatment on large segments of populations throughout the world; that is, biomedical ideas have become part of health cultures. Medicalization can be seen in many forms, from the overprescription of anti-depressant

Box 4.1 The underdevelopment of health

It is evident from many studies that efforts to 'develop' the so-called developing countries are closely tied to the health of populations in those countries. One might assume that development would improve the health status of a country, but this is not always the case. In a landmark review three decades ago, Charles Hughes and John Hunter (1970) showed that many specific development efforts had unanticipated detrimental side effects. A famous example is that building dams on rivers created still water and thus produced conditions in which the snail vector that was an essential part of the life cycle of the schistosomiasis worm could thrive. Surveys showed that the incidence of schistosomiasis increased dramatically along the reservoirs and irrigation canals that resulted from building dams.

In the mid-1980s, Robert Stock (1986) found that little heed had been paid to the lessons of the Hughes and Hunter study by either planners or researchers. Also, he suggested, the more basic problem was not development, but *underdevelopment*, an integral part of the worldwide capitalist system. Vicente Navarro (1974) argued that capitalism created an uneven distribution of resources between developed and developing areas by stifling the diffusion of technological and cultural innovations, limiting capital flow, and creating dual economies. In her work, Meredith Turshen (1977) demonstrated that disease burdens in developing countries such as Tanzania could be traced ultimately to political and economic decisions related to capitalism. Focusing on Africa, Stock (1986) listed several effects of underdevelopment on health: (1) emphasis on curative rather than preventive medicine; (2) an extreme urban bias in the distribution of health care resources; (3) disparities in access based on class; and (4) shortage of essential drugs and equipment. These problems were evident in the days of colonial rule and persist in the post-colonial period.

drugs to the popularity of television shows like *ER* and *Chicago Hope* to the promotion of high-tech medical equipment. Medicalization may pervert or distort reality: thus drug use often masks the true social and economic conditions that lie behind problems like depression, and medical soaps glorify doctors and technology and create unrealistic expectations (Turow 1997). Several scholars have written about the way medicalization can be used as a method of social control by dominant groups (e.g. Zola 1972). This can be seen in the power doctors have to diagnose illnesses and prescribe treatments, and in the way people are labelled as having various stigmatizing mental illnesses. At the same time, medicalization creates unwanted needs and ensures greater profits for companies that produce health care resources. What makes medicalization particularly effective is the way in which it creates ideologies about biomedicine and health care that often become 'understood truths' or 'common sense', part of our cultural belief system (Krause 1977). Biomedical ideologies stress the biological as opposed to the social, cultural, political and economic causes of disease; reinforce the hegemony of the medical 'expert'; and create widespread acceptance of disease classifications that may not have local cultural relevance.

In recent years, much has been made of the very widespread moves within developed countries to *restructure* their health care delivery systems, most notably by making massive switches from public to privatized care. Scarpaci (1989b: 2) states that '[there is] growing consensus worldwide that the unfettered marketplace effectively regulates the availability and accessibility of health services'. He also argues, however, that participation is not just a response to economic crises or that there is a global conspiracy to roll back the welfare state. Rather, one must consider that restructuring in any place involves a dynamic interplay between governments, private businesses, capital and health care consumers. Many arguments have been made for or against privatization; we mention only a few here. Proponents often claim that putting health care into private hands increases efficiency and responds more effectively to consumer needs. They feel that the World Health Organization's goal of achieving complete physical, mental and emotional well-being puts unnecessary strain on the public purse and that government health bureaucracies are very wasteful of taxpayers' money. Furthermore, privatization is said to provide individual liberties and foster innovation. Those opposed to privatization claim that the state should intervene in health care as part of its social contract with its people. If one tries to eliminate the welfare state, social relations will break down. Public administration of health care benefits from central planning and promotes non-economic benefits such as altruism and social cohesion. The debate will no doubt continue for a very long time.

Attempts to dominate the health care field, fortunately, are often met with *resistance*. One way to enact resistance is to expose the flaws of biomedicine and the intentions of health care entrepreneurs (Illich 1976; Navarro 1974) (Box 4.2). Consumers can be encouraged to question the diagnoses and prescriptions of their doctors, to investigate alternatives to biomedicine, and to ask who profits from health care provision. Currently we are witnessing movements, by women in particular, to 'take control of their own bodies', but this idea can be traced back at least to the nineteenth century, when the hydrotherapy movement empowered women by demystifying such life events as menstruation and childbearing and treated them as natural occurrences rather than illnesses (Cayleff 1988).

Box 4.2 Limits to medicine

Many causes are championed by an outstanding personality. Ivan Illich (who once worked for five years as a parish priest in an Irish–Puerto Rican neighbourhood of New York City) is well known for his leadership of the resistance movement against the biomedical establishment. Illich (1976: 3) has simply declared that 'the medical establishment has become a major threat to health', a condition that he terms *iatrogenesis* or health system-generated disease. Medicalization, Illich claims, has led to imposing so-called miracle cures on patients that are later found to be dangerous. It is also well known that hospitals and other health facilities may be sources of infection. The medical system creates problems by promoting expensive and often unnecessary technologies and costly medicines when cheaper generics are just as effective. Medicine also creates relations of power in which health practitioners dominate consumers. And the medical system transforms 'pain impairment and death from a personal challenge into a technical problem' (p. 10). The solution to this 'medical nemesis', Illich argues, is to limit the power of medical personnel and place the control of the health of the public in the hands of lay people. In other words, Illich advocates a grass-roots, bottom-up approach to health care that focuses on those who experience health and disease in their everyday lives.

Two examples will provide more detail on the topics just discussed, namely inequalities, medicalization, privatization and resistance. The first concerns the worldwide domination of the legal drug trade by multinational pharmaceutical companies (MPCs), private companies that largely control the manufacture, distribution and sale of medicines all over the globe (Gesler 1989, 1994). These private corporations are the epitome of the profit motive inherent in capitalism. Since they promote self-care and push sales to private practitioners, they tend to expand the privatization of health care. Marketing the products of biomedical research, they medicalize by pushing alternatives (e.g. efficacious herbal remedies) off the shelves of pharmacies and patent medicine stores. Advertising creates an ideology that only expensive brand-name drugs will do. MPCs have come to dominate medical markets because they can make and sell legal drugs more efficiently than public health care programmes can. Box 4.3 is a case study of a specific drug marketed by MPCs.

Although drug companies do provide a much-needed and relatively efficient service, their hegemony creates many problems, especially in the less developed world: they control what drugs are distributed and therefore what diseases can be treated; they dump old drugs and those with questionable efficacy and known side effects; they control research and development; and they promote over-the-counter as opposed to prescription sales. They also promote inequalities in the distribution of drugs as the poor and those who live in either rural areas or deprived areas of cities have less financial and geographic access to them. It is extremely difficult to resist the hegemony of MPCs, although some national governments and international organizations have tried. India, Cuba, Sri Lanka and Mozambique have attempted either to manufacture their own drugs or to control the types of drugs coming in from outside, and United Nations organizations have tried to make the MPCs comply with lists of essential drugs, but the dominance/subservience relationship persists.

The second example concerns the (re)emergence of alternative or complementary medicine in the US. A report in the *New England Journal of Medicine* (Eisenberg *et al.* 1993) which stated that one-third of the adult US population used some type of 'unconventional' therapy confirmed and legitimized what many people knew was happening: people in increasing numbers were 'voting with

---

Box 4.3 The case of Depo-Provera

Depo-Provera is an injectible contraceptive manufactured by the Upjohn Corporation for use by women. It suppresses ovulation for from three months to a year and by the mid-1980s had been given to over ten million women in eighty countries, mainly in Asia (Duggan 1986). It has several potential advantages, including its ease of administration and long-lasting effects. It is invisible, and thus escapes the notice of men who do not want their wives to use any type of contraceptive. Advocates say that it is the best way to prevent women from dying from abortions performed by local midwives.

For many, however, the disadvantages of Depo-Provera far outweigh its advantages. The fact that the Federal Food and Drug Administration ruled in 1972 that the injection could not be promoted for contraceptive use in the USA (although it was licensed for other limited purposes) and the fact that the Swedish International Development Agency refused to supply it to countries that requested it should be clues as to its dangers. Studies of animals showed that it could produce breast and endometrial cancers, increase risk of artherosclerosis and osteoporosis, lower life expectancy, and lower resistance to infection. Women implanted with Depo-Provera may experience depression, hair loss, headaches, weight loss or gain, and other side effects such as losing interest in sex. These complications clearly

---

affect a woman's social and psychological as well as her physiological well-being. Furthermore, the contraceptive may mask key indicators (e.g. irregular bleeding) by which a woman can monitor the true condition of her body. In other words, Depo-Provera 'inscribes' itself on, and leads to lack of control over, one's body.

Depo-Provera helps us to understand the difference between birth control and population control (Duggan 1986). It may be a useful choice to a woman who is practising birth control; that is, deciding when or whether to conceive a child. On the other hand, it may be used by advocates of population control who seek to lower fertility rates. For those population control promoters who believe that women in developing countries are too ignorant or irresponsible to use other means of contraception such as pills, Depo-Provera is an ideal contraceptive.

The Depo-Provera case stimulated a debate that has no clear-cut solution. Advocates accused the Swedish government of exercising a neo-colonial form of power in preventing its use where it was desperately needed. Some critics wanted to ban it entirely. Others promoted its use under 'ideal' conditions in which a woman could be closely monitored, a situation that is highly unlikely for most women.

their feet' and seeking alternatives to biomedical treatments, resisting to some extent those who medicalized their lives (Gesler and Gordon 1998). Although alternatives such as homoeopathy provided serious opposition for biomedicine in the nineteenth century, the discovery of germ theory and strong measures taken by the American biomedical establishment to legitimize itself and marginalize others led to the relative decline of these practices. Now, people are turning again to alternatives for at least four reasons. First, there is dissatisfaction with biomedicine because of its uncaring attitude, inability to treat chronic illnesses, the serious side effects of medicines, high costs, and inequalities in resource distributions. Second, the 'baby boom' generation is turning 50 and demanding more health care. Influenced by the changes that took place in the 1960s as they were growing up, these people are more likely than others to look for non-biomedical solutions. The third reason is that in a quest for overall wellness, many find more satisfaction from practitioners who spend more time with them and help them to help themselves. Finally, it seems that many people are becoming more pragmatic in their health care choices, more willing to shop around, more willing to take alternatives into their health care culture.

One might think that increased use of alternative or complementary practices would lead to more equality in access to care. However, evidence to date on this issue is equivocal. For example, state laws that regulate alternative practices vary (that is, they are more or less strict in terms of scope of practice, educational requirements and licensing requirements), and this has an impact on the distribution and availability of alternative practices (Baer and Good 1998). Although insurance companies are increasingly covering non-biomedical care and thus potentially increasing equality, people still pay around 75 per cent of alternative therapy costs, and thus the more wealthy are favoured (Gordon and Silverstein 1998). And one can also discover geographic inequalities in the distribution of alternative practices: a study carried out in Washington, Oregon and California showed that the highest alternative practitioner-to-population ratios were in high-amenity, non-metropolitan urbanized areas as opposed to metropolitan or rural places (Osborn 1998).

Unhealthy societies

The structural approach deals with, among other things, social forces that lead to economic inequalities. There is a great deal of evidence to show that material deprivation, measured, as examples, by poverty levels or quality of housing, is associated with poor health (see, for example, Townsend et al. 1985). A result, and symptom, of deprivation is homelessness, especially prevalent in large cities (Figures 4.1 and 4.2). One can easily map out areas where people are deprived, as Knox (1995) has done for urban areas. We note, however, that there has been a poverty of deprivation studies in health care because they fail to take into account relative levels of deprivation within groups of people.

Recently, a very striking finding has come to light in developed countries, namely that where income differences are smaller within countries, health indicators such as mortality and morbidity rates and life expectancy are better (Wilkinson 1996). For example, a study of eighteen industrialized countries from 1950 to 1985 showed that 'Income inequality and relative poverty rates appear to be of greater importance for the variation in infant mortality rates than the level of economic development between rich countries' (Wennemo 1993: 429). In another study, Andes (1989) found that differentials in infant mortality between two Peruvian communities were due mainly to economic diversity, income disparity, social class fluidity, women's autonomy, and some refined measures of medical care and public health resources.

Wilkinson (1990) notes that as countries gain more wealth, the link between wealth and health weakens and income distribution becomes more important. It appears that once a society reaches a certain overall level of economic growth, what matters most is not absolute income but relative income levels. This means, in effect, that the social forces that underlie income disparities (class, racism,

Figure 4.1 A homeless person in Washington, DC. The irony of this picture is, of course, that justice is hardly being served if someone like this is sleeping out in the US capital city. Photo by David Carr.

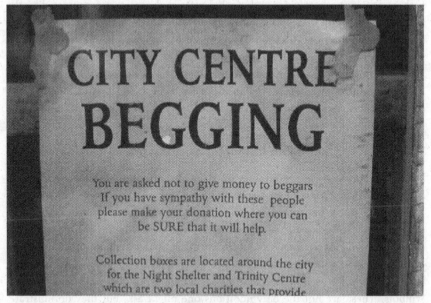

Figure 4.2 Signage discouraging homeless persons from central Winchester. Photo by Robin Kearns.

gender), rather than material factors, are the limiting factors to the promotion of good health. Although economic growth in quantitative terms is important, qualitative changes that improve the social fabric and distributional justice are more important to health.

Researchers such as Wilkinson and others are saying that what makes modern societies unhealthy is structural factors. What matter most are things such as social position, fears about unemployment, and the extent to which there is social mobility. In the past, at the beginning of the epidemiologic transition, when contagious and infectious diseases were the leading causes of illness and death, biomedicine could identify single risk factors for single diseases. Now that the so-called 'diseases of affluence' or degenerative diseases such as heart problems and cancers are the biggest killers, they have, ironically, become mainly the diseases of the poor.

Why should relative deprivation be so important? Why do societies that keep the gap between rich and poor relatively narrow have higher levels of well-being than those in which there are wide gaps in standards of living? Wilkinson (1996) suggests that egalitarian societies share a common characteristic: social cohesion. Here the values of individualism and the competitive marketplace are restrained by a sense of social morality. There are higher levels of social capital and less antisocial aggressiveness. Although all causes of death in developed economies are higher where there are greater income discrepancies, it is the 'social diseases' such as accidents, homicides and alcohol-related deaths that are most directly affected by a lack of social cohesion or social responsibility. We have very good evidence that psychosocial stress, often caused by social fragmentation or lack of social morality, has a strong influence on morbidity and mortality rates. Thus the ability of people to tap into supportive social networks becomes extremely important. These kinds of resources are more likely to be found in socially cohesive, caring societies.

The humanist approach

From the humanist perspective, what is most important about a situation is an individual's subjective experiences. How do people feel about the activities they engage in? What meanings are there in symbolic landscapes? Do people experience a strong sense of place where they live? In the realm of health care, the humanist approach leads us to ask questions about (1) the beliefs of different actors in medical encounters; (2) feelings of identity with places where health care is administered; and (3) the role of concrete and abstract symbols. These are not the only questions one might ask, but they do illustrate well the 'meaning-centred' or human agency way of looking at health care (Good 1994).

When someone is being treated for a medical problem, several actors may be involved: the patient, nurses, doctors, lab technicians, family and friends, and so on. A very important question for the progress of the illness episode and its outcome is what the different actors believe about the particular problem and the best way to treat it. Apart from the very difficult question of whether *beliefs* are 'right' or not, one has to face the issue of whether the ideas of the different participants in a medical situation are in tune with each other (Fabrega 1980). One would expect more positive outcomes if beliefs were congruent, if everyone involved shared the same health culture. In contrast, negative outcomes are more likely when beliefs clash. Take the example of a medical encounter between someone who believes that their disease was brought about by witchcraft and a Western-trained doctor who believes in germ theory. Or think of the difficulty you might have when talking to your doctor: you think in terms of your symptoms and bits and pieces of information about your condition you have picked up from various sources and the doctor comes on strong with medical jargon that is the result of years of training but has little meaning for you.

It is useful when considering health beliefs to think in terms of the explanatory models (EMs) that were first introduced in Chapter 2 (Kleinman 1978, 1980). These are the ideas or beliefs that people use to make sense of their illnesses and to evaluate possible treatments. They consist of both conscious knowledge and 'understood truths' that people may not be consciously aware of. Each person's EM is unique, but people within a culture group tend to share similar specific beliefs. EMs are often focused on the significance of an illness to those with whom one interacts, such as family, friends and workmates. It cannot be overemphasized that the EMs of patients and health providers may not be in agreement at many points, despite the extent to which large segments of populations have become medicalized (that is, enculturated with the terminology and ideologies of biomedicine).

As an illustration of EMs, take the case of diabetes. A great deal is known about the risk factors for this disease (e.g. genetic predisposition, obesity and lifestyle) and what people who have or are at risk of having diabetes should do to treat or prevent it (e.g. low-sugar diets, exercise). However, unless one understands what people's EMs related to diabetes are, it is extremely difficult to educate people to 'do the right thing' about this chronic illness. For example, Cohen *et al.* (1994) looked at the EMs of Midwestern European Americans being treated for diabetes and found that they did not know the cause of their diabetes, giving instead unusual responses such as 'a car accident, having a "moody personality", an episode of the flu, or eating foods foreign to one's system' (p. 61). Most people thought the disease was permanent; they were concerned about its complications (e.g. amputation of limbs, blindness); and said the most important difficulties they faced were dramatic changes in lifestyle and personal relationships. Furthermore, patient EMs differed from those of the health professionals who were treating them. Cohen *et al.* (1994) and Hunt *et al.* (1998) have found that diabetes EMs are shared among people with a common cultural background and that people pick up ideas about the disease from family and friends.

If a person identifies in a positive way with the *place* in which care is being provided, indications are that the treatment will be more successful. For example, it has been shown that it is beneficial to alcoholics to have a feeling of *rootedness* in places where they live or go to for help (Godkin 1980). It is important for health care personnel to have positive feelings about the places in which they work as well. In the USA, a major problem in rural areas is physician retention, or inducing doctors to serve communities under difficult conditions, including isolation from colleagues, lack of amenities, and long working hours with relatively low pay. Most studies of retention have focused on the degree of turnover or specific factors related to how long doctors stay in rural areas. Malcolm Cutchin (1997) went beyond these types of research, taking a humanistic (or, more specifically, a pragmatist) approach to examine the *process* leading to retention. He calls his theoretical perspective *experiential place integration*: the idea is that 'integration is a type of progress that builds bonds with place, that in turn encourages retention' (p. 28). Cutchin conducted in-depth, open-ended interviews with fourteen physicians in Appalachian Kentucky who had successfully integrated into rural communities. Analysis of the interviews showed that place integration occurred within three domains: self, medical community and community-at-large. A key finding was that integration could be characterized by three dimensions: (1) security (e.g. feelings of safety, stability and confidence); (2) freedom (e.g. ability to act out one's desires and achieve one's goals); and (3) identity (e.g. coherence of self with others and the environment).

Because *symbols* are so much a part of our everyday lives, we seldom think about what they are or what they mean to us. But, whether consciously or unconsciously taken in, both concrete and abstract symbols can be very powerful. When in a hospital, for example, one is surrounded by such symbols as a physician wearing a white coat that could mean honesty or purity, high-tech machinery that might stand for competence or create fear; or stories told by fellow patients that foster myths about treatment and thereby engender either confidence or distrust. Symbols provide their meaning because we relate them to other aspects of our daily lives. Thus the white colour of a physician's coat may remind us of a freshly painted house, pure water or a sunny day.

Many more examples of symbols in health could be given; however, since we will take up the whole issue of symbols and symbolic landscapes in Chapter 7, let one more illustration suffice here. In north-western Australia, the aboriginal Murngin people have developed an elaborate set of symbols, based on images of body destruction, which they convert through rituals to feelings of health (Munn

1969). What gives these symbols power is that they are connected to many other aspects of Murngin life besides health, including menstruation in women and circumcision in men, wet and dry seasons, food plenty and food scarcity, snakes and other animals, and floods and tides. In other words, symbols related to health are interconnected with symbols related to other aspects of daily life through semantic networks.

Deprived people

We can further illustrate the humanist approach by returning to the topic of deprivation studies. Earlier in the chapter, we argued that, at least in developed economies, absolute deprivation was not as important to health as relative deprivation. We introduce here a second aspect to the poverty of deprivation studies, and that is that the manipulation of census or survey data to depict geographic areas of deprivation fails to address adequately the beliefs and experiences of the deprived themselves. Thus, although we fully acknowledge the importance of social or structural forces in determining overall differences in levels of health among modern societies, we also point to the need to get nearer to people themselves to understand the personal effects of deprivation.

Surely one of the best attempts to enter into the lives of the deprived is Jocelyn Cornwell's (1984) study of the beliefs and everyday practices of a group of two dozen people in the west Bethnal Green area of the London Borough of Tower Hamlets, a place notorious for its deplorable living conditions. Despite a mythology that the residents of Bethnal Green developed (and local authorities encouraged) about living in a community of cheerful, cooperative people, life in council (low-income) housing in a city such as London is truly hard-earned.

In getting to know her study subjects, Cornwell discovered that they often spoke in two different 'voices' or used two different types of discourse (see Chapter 5 for more on the importance of language). At first, unsure of their relationship with the researchers, people used *public accounts* or 'sets of meaning in common social currency that reproduce and legitimate the assumptions taken for granted about the nature of social reality' (Cornwell: 1984: 199). When she had gained their confidence, people used *private accounts* that 'spring directly from personal experience and from the thoughts and feelings accompanying it' (p. 199). Both accounts are important; the public one reveals what a person thinks the world would like to hear and the private one speaks to what the person really experiences and feels.

Cornwell's subjects talked about many things, including their work experiences and family relationships; here we will concentrate on what they said about health. People's public discourse on illness dwelled on the idea that one had to be seen to take the proper moral stance with respect to problems such as heart disease or cancer; one had to take illness seriously and not be seen to be malingering. In private discourse, however, people revealed personal fears and actual behaviour. What kept many people hard at work, for example, was a stoical approach to life, but also a realization that too many sick-days could lead to loss of a job. It was acceptable in this culture of health for men to come home for a rest for a few days, but women were expected to keep going.

Residents of Bethnal Green had developed a health culture that blended biomedical ideas and a traditional, common-sense moral approach. When asked about the causes of illness, their public speech included things such as stress, germs and viruses. Private speech, in contrast, revealed the importance of bad drinking habits, family squabbles, or the loss of a loved one. Publicly, people were respectful of health care personnel; privately, individual practitioners were criticized. The local culture also created a hierarchy of respect for health care resources: hospitals, which were said to treat 'real' diseases like cancer, were held in the highest esteem; doctors, who dealt with 'normal' diseases, came next; last came community service personnel (e.g. school health nurses and health visitors), who were thought to be simply doing jobs that required ordinary common sense.

Studies like Cornwell's reveal a great deal about the people who are supposed to be the primary beneficiaries of a health care system. What is perhaps most striking is a kind of paradox that their discourses expose. On the one hand (and this comes out mainly in private discourse), they sense that something is wrong, that they are being treated unfairly, that the system gives them little respect. On the other hand, as Lemert (1997b) notes, victims of deprivation often praise those in authority, blame themselves for their plight, and sink into a fatalism in which they know for certain that they are going to mess up their lives. Thus they become caught up in an endless cycle of deprivation and stoically suffer the consequences.

Integrating structure and agency

Both the structuralist and the humanist approaches have proved to be very useful tools for analysing health care in places. However, both perspectives have been criticized for leaving out what the other provides; that is, structuralists are faulted for not paying attention to individual actions and experiences, and humanists are faulted for failing to recognize the importance of underlying social forces. Given these criticisms, it is not surprising that scholars have attempted to blend the two approaches into a more comprehensive way of looking at the world. The idea is that 'adequate explanations of human actions must include both the fatalism of social structures and the creative spontaneity of the lifeworld' (Ley 1981: 42). The result of linking these two perspectives has been called several things, including Marxist humanism and structuration. These blends try to reconcile a series of seeming opposites: structure and agency, society and individual, hegemony and self-expression, cultural norms and personal biographies, social systems and everyday practice. The important thing is that the dichotomous elements are continually interacting with each other; they are certainly important on their own, but usually more so in relation to their opposite. Indeed, it is probably better not to think in terms of dichotomies at all. A very important aspect of the new cultural geography is that it focuses on the dialectic between structure and agency as it is worked out in specific places. As Cosgrove and Jackson (1987: 99) state, 'particular cultural forms can be related to specific material circumstances in particular localities on the ground'.

One attempt to link structure and agency is called *structuration theory* (another is *critical realism*, see Box 4.4). The basic ideas are attributed to the sociologist Anthony Giddens, but several other scholars, including geographers, have made significant contributions (see Giddens 1976; Gregory 1981; Johnston 1986b; Pred 1984). Structuration theory tries to balance the structures of society that enable and constrain human actions with the freedom people have to make choices within structures, and even to change structures. Human activities and values are interpreted from both the viewpoint of individual actors in a situation and the social relations out of which actions and values arise. In Gregory's words,

> in the reproduction of social *life* (through systems of interaction) actors routinely draw upon interpretive schemes, resources and norms which

---

Box 4.4 Critical realism

Structuration is discussed at some length in this chapter as a guide to incorporating both structure and agency into health care geography. Another approach is through *critical realism*. The major work on critical realism comes from Anthony Sayer (1992). Although Sayer's work touches on several important concepts, including the issues of causality, open and closed systems, and necessary and contingent relations, we focus here on the realist trilogy of events, mechanisms and structures. An *event* is an instance of an individual or group action (e.g. someone snubs you at a party). An event occurs because it is triggered by a *mechanism*, which is a causal power or way of acting (e.g. going to parties). *Structures* include the underlying forces discussed earlier in this chapter, such as race and class. They can be thought of as contexts for human actions. Thus the reason you were snubbed may stem from class differences.

---

are made available by existing structures of significance, domination and legitimization, and . . . in doing so they immediately and necessarily reconstitute those *structures*.

(Gregory 1981: 9–10)

What is extremely interesting about these concepts for geographers is that agency and structure interact over space and time, an idea that is incorporated into the notion of time geography (Box 4.5).

Box 4.5 Time geography

An exciting development for geographers is that they have been able to contribute a great deal to structuration because the theory has been examined empirically by seeing how it works out in specific times and places. Particularly valuable has been the development of *time geography* by Torsten Hagerstrand (1982), which is based upon components of people's everyday activities, their *time–space routines*. Path and project are two basic concepts used in time geography. A *path* is the route someone takes through space and time as he or she carries out an activity. A *project* consists of the tasks necessary to satisfy a goal and may involve several intersecting paths that are linked with the required resources to carry it out. Paths and projects are constrained by various social and environmental factors (structures), but are also governed by individual desires and actions (agency).

A very simple example shows how this all works. Suppose your project is to have a medical check-up by your family doctor. We can trace your path as you make your way from home or work to the doctor's office, wait there, have your examination, and move on. We can also look at the paths of others involved – the doctor, a nurse, a lab technician – as well as consider the equipment that needs to be available for the check-up. Your path and those of the others is constrained by many things such as finding a time when everyone can come together, modes of transportation available, financial considerations, and availability of medical supplies. At the same time, all the actors have some freedom to choose when, where and how the required activities are carried out (for example, the doctor may keep you waiting because he or she wants to do something else first or you decide to cancel the appointment because of something that came up at home or work).

For a further illustration of time geography, let us return to the example of diabetes that was used earlier in the chapter to illustrate the concept of explanatory models or EMs. Examining EMs gets at the beliefs that people

---

have about what causes diabetes and how it should be prevented or treated. But where do people obtain this information? To examine this geographical question, a team of researchers (Cravey *et al.* 2001) developed the concept of socio-spatial knowledge networks (SSKNs). These are maps of where people pick up ideas about diabetes over time; they depict locations of people, institutions and places where people obtain information and their paths around these locations as they carry out projects that increase their knowledge about diabetes. These maps can be enhanced by including such items as frequency of contact with locations, quality of information obtained, and comparison with maps of people's activity spaces. Knowing where people obtain or would like to obtain information on diabetes leads to the identification of potential sites for interventions.

Baer *et al.* (2000) combined concepts from both structuration theory and time geography to look at the critical questions of physician recruitment and retention. Using as an analogy the shape of a wineglass, they show how one can graphically display three phases of a physician's career: (1) study at a particular institution (the wineglass stem); (2) recruitment from an area around the institution (the base); and (3) diffusion outwards over space and time from the stem to take up jobs (the bowl). Structures in this illustration include requirements of the medical profession, governmental laws and regulations, liability insurance, and state of the overall economy; agency includes decisions about where to go to school, what kind of job to pursue, and places to move to. Although the data for studies like this is hard to find and analyse, the *wineglass model* enables one to gain an overall picture of a physician's life course over space and time.

Leonard Baer (2001) was interested in the question of how well international medical graduates (IMGs; doctors working in the USA who received their medical degrees outside the USA or Canada) were accepted in two rural communities in the south-eastern part of the country. He decided that critical realism was a good way to come to grips with this rather nebulous problem that could not satisfactorily be examined by, for example, quoting, or mapping statistics on how many IMGs actually practised in different regions of the USA. Rather, he wanted to examine the *process* of achieving acceptance. To accomplish this task, around 125 people were interviewed in the two places, including IMGs, US-trained physicians, other health personnel, patients and members of the community.

Careful analysis of tape-recorded interviews revealed sets of events, mechanisms and structures/contexts. The main event was acceptability, which was theorized in terms of creating the 'other' or 'geographies of exclusion' (see Chapter 6). Mechanisms included the role of a hospital in a place and the role of pioneer IMGs in gaining acceptance. Two of the main structures encountered were race and gender. Although critical realism helped to provide a theoretical framework for the study, two further concepts were developed from the analysis. One was the importance of the identity of the two places. That is, the communities created environments in which structures, mechanisms and events played themselves out in different ways (that is, place mattered). Furthermore, it was discovered that different types of boundaries were very blurred. For example, it was hard to say who was perceived by townsfolk to be an American or who was thought to be a foreigner. What community did an IMG from India really belong to if she worked with colleagues in a hospital, practised Hinduism with other Indians, shopped in the local mall, and kept close contacts with relatives in New Delhi?

Structuration, time geography and critical realism are guides to research rather than philosophical approaches such as humanism and structuralism. In practice, it is quite difficult to determine what factors should be labelled either 'structure' or 'agency', or to show how various structures and agents interact. Most health studies look at what is happening at only one period in time, whereas structuration and time geography really call for diachronic investigations like the one carried out by Baer *et al.* (2000). However, these approaches or methodologies have proved their usefulness, as we hope to show with the following examples from the literature on mental health.

Treating the mad

Health geographers have a distinguished tradition of research into mental health (see, for example, Chris Smith and John Giggs's edited book *Location and Stigma* (1988) and the special issue of *Health and Place* (1997, vol. 3(2)) titled 'Space, place, and the asylum'). More recently, another special issue of the same journal focused on post-asylum geographers (2000, vol. 6(3)) (Box 4.6). Geographic work on mental illness provides excellent examples of the integration of structural and humanist approaches.

Box 4.6 Post-asylum geographies

It seems to be difficult to know what to do with people who have mental health problems. Assuming one can define what mental illness is, the problem becomes one of treating people with mental health problems and finding places for many of them to live. In the nineteenth and twentieth centuries in industrialized countries, people with mental health problems

---

were often housed in large institutions or asylums. Towards the end of the twentieth century a move was made to take people out of asylums and put them into the community. This strategy has failed in many places, often sending those with mental health problems onto the streets or into prisons. Recently, a concern with the 'risk' and 'danger' perceived from violent patients living in communities has led to talk of confinement again (Philo 2000). Thus treatment for those with mental illness moves through yet another cycle (see also Geores and Gesler 1999).

Jennifer Wolch and Chris Philo (2000) have traced geographic work in the post-asylum era in terms of three waves. From the mid-1970s into the 1980s, during the first wave, most health geographers focused on questions of the spatial distribution of people with mental health problems and the places that treated them. Researchers traced people as they were moved out of asylums, into communities (where they often became concentrated into urban ghettos), and into the ranks of the homeless. Second-wave mental health geographers, informed by social theory, worked at smaller scales of analysis and dealt with questions of disability, identity and place. An earlier focus on deviance was replaced by a concern for difference. Research tended to ignore structural and institutional concerns, as well as policy issues. Wolch and Philo are disturbed by this latter research turn and have called for a third wave that would move on to studies that combine what goes on in places and wider political, economic and social structures into work that informs policy. They suggest that mental health geographers work from a *(dis)placement model* of service delivery in which the poor, including those with mental health problems and the homeless, are constantly shunted around from one unhealthy place to another.

Perhaps the most important event for the geography of mental health since the 1960s in Western countries has been the deinstitutionalization of care, the movement of large proportions of those with mental illness from large institutions to smaller facilities or homes throughout communities. In many cities the process of deinstitutionalization created places where the mentally ill tended to end up, areas that Dear and Wolch (1987) call the *service-dependent ghetto* or *landscapes of despair*. Three interrelated institutions or processes that we will think of here as structures had a great deal to do with the formation of these ghettos. The *capitalist economic system* produced conditions of poverty and unemployment; the deinstitutionalized mentally ill in particular suffered from these conditions. Changing economic circumstances also led to processes of *urbanization*. Of particular interest here was the creation of *zones of transition* that formed at the edges of expanding central city commercial zones through two processes:

(1) abandonment by those who could leave and move out to suburban areas; and (2) takeover by the relatively poor. These areas became service-dependent ghettos. The third 'structure' was the *welfare state*, closely linked, again, with the capitalist system. Although it seldom sought to change the underlying social forces that created poverty, the welfare state nonetheless acted as the conscience of a nation by assuming responsibility for deprived and ailing groups such as the mentally ill. It was welfare state institutions that, beginning in the 1960s, decided that large facilities were no longer treating patients adequately and that new techniques such as drug therapies pointed towards deinstitutionalization. As a result, many people with mental illness drifted to zones of transition.

What agents were involved in this process? *Health care professionals* – psychiatrists, nurses, asylum directors and others – had a great deal of control over such matters as who was declared to be mentally ill, what hospitals should close and who should be moved. *Mental patients* themselves had to make decisions within the constraints imposed by the various structures described above. Some tried to make their own way, but their stigmatized status often prevented them from obtaining jobs or being accepted in 'respectable' communities. *Urban residents* who resisted the placement of new, smaller mental health facilities in their communities were active agents in forcing many people with mental illness into ghettos. *Land use planners* also became involved when they zoned areas of the city in ways that either encouraged or discouraged the location of mental health facilities.

It is difficult to separate structures and agents in this situation. For example, land use planners are very much a part of the urbanization process and mental health professionals are employees of the welfare state. What is more important than sorting out structure and agency, however, is thinking through interactions between the various factors that create landscapes of despair. Urbanization takes place inexorably, it seems, and yet people like the mentally ill also have some say in where they live, and they certainly change the character of the areas to which they migrate. Many mental health professionals disagreed with the decisions made by welfare state institutions and fought to provide better care for their clients.

Hester Parr (1999) adds a new dimension to the question of what happens to those with mental illness by focusing on their 'projects' to find therapeutic spaces or places in Nottingham, England. Feeling marginalized and isolated, as well as being poor, people with mental problems trace paths through the city, seeking spots with which they can identify and where they feel secure and comfortable. Their movements are severely restricted by the structures of society, probably more so than for most of us. For many, mental health facilities are places to be shunned because there they are stigmatized by the labels of biomedicine, their personalities are taken over, and they are made to feel impure and different. To escape the 'psychiatric gaze', many seek out public places in which to express themselves. In these places, however, they often come into contact with the structures within society that discourage 'abnormal' behaviour; thus they are subject to, at the least, disapproval, and, at the worst, arrest. Still, those with mental problems do manage to find therapeutic places such as cafés, parks, city squares and street benches that they can claim for a time as their own.

Some of those with mental problems have, fortunately, found a way to resist the structural forces that impinge on their freedom to express themselves and find healing places. Through collective action they can create, in a sense, their own structures. Collective action is particularly difficult for this group, however, because mental illness is usually held to be an individual problem and because the mentally ill are thought to have little political will or desire to help themselves and others. Indeed, the most influential organizations that advocate for them have been *for* patients rather than *of* patients. This situation is changing in the UK, where, for example, a patient group has formed in Nottingham. Patients in these groups have taken on projects of helping people to find new therapeutic settings in which to live; changing the health cultures within formal service facilities; and changing medical conceptions about mental illness and its treatment. In one instance in Nottingham, patients attempted (unsuccessfully) to prevent the closure of an asylum because for many it had become a place of sanctuary, despite the 'bad name' it had in the community. What is needed, Gleeson and Kearns (2001) argue, is 'inclusive service landscapes that cherish the diverse values and interests of carers and care recipients'.

Conclusions

This chapter can be taken as an attempt to lure the reader away from a traditional medical geography focus on disease and biomedical treatment to views of health care that reveal both society-wide forces that constrain and provide opportunities for better health and the personal experiences and beliefs of people who seek and receive health care. This is not to say that the work that medical geographers have done in the past on the ecology of specific diseases or the location of and access to health care personnel and facilities is not important. It is, but so are the social contexts in which health care takes place and the biographies of those who have the most to gain or lose from its delivery.

Arguments were presented for the benefits of using both structuralist and humanist approaches. Advocates of one or the other stance still extol their virtues as health geography moves (inevitably, we feel) to post-positivist positions. As the latter part of the chapter argues, however, it is also useful to keep structure and agency in tension, to allow for each, and attempt to blend them, guided by an approach such as structuration. One possible way to do this is by creating models that operate at different levels. Ann Millard (1994) has proposed such a model for examining the causes of high rates of child mortality that roughly parallels biomedical/positivist, humanist and structural concerns. Here is her description of the model (p. 253):

> The proximate tier includes the immediate biomedical conditions that result in death, typically involving interactions of malnutrition and infection. The intermediate tier includes child care practices and other behavior that increase the exposure of children to causes of death on the proximate tier. The ultimate tier encompasses the broad social, economic, and cultural processes and structures that lead to the differential distribution of basic necessities, especially food, shelter, and sanitation.

Any one study could examine a single one of these three levels or tiers, but all are important, and a thorough examination of such a phenomenon as child mortality requires a synthesis of all three (Gesler *et al*. 1997).

Further reading

Cornwell, J. (1984) *Hard-Earned Lives*, London: Tavistock. This is a classic study of how disease and health care interweave with the daily lives of inner-city London residents. Cornwell treats her study subjects with an honest and sympathetic eye and exemplifies the ethnographic method of qualitative analysis.

Dear, M. and Wolch, J. R. (1987) *Landscapes of Despair: From Deinstitutionalization to Homelessness*, Princeton, NJ: Princeton University Press. This book by veterans of the 'deinstitutionalization wars' examines the ways in which structure and agency interact to create deprived or unhealthy spaces for people living in urban areas.

Scarpaci, J. L. (1989) *Health Services Privatization in Industrial Societies*, New Brunswick, NJ: Rutgers University Press. This collection of chapters by health geographers deals with various forms of the process of health 'restructuring' that have taken place in Western societies.

Smith, C. and Giggs, J. (1988) *Location and Stigma: Contemporary Perspectives on Mental Health and Mental Health Care*, London: Unwin Hyman. This edited volume adds to the distinguished collection of work that health geographers have produced on people with mental illness.

5

LANGUAGE/METAPHOR/HEALTH

Introduction

A young woman, wearing a flimsy paper gown and sitting in a chilly, sterile hospital examination cubicle, has just been told by an oncologist that she has breast cancer. She is frightened and very upset, worried about the pain she will have to endure and the changes this will make to her life. She pours out her fears in a torrent of words, but the doctor cuts her short, tells her the main thing is to attack the cancer head on, and goes on at great length to describe to the uncomprehending woman the technical details of the course of chemotherapy he recommends. What is he talking about, she wonders; what was she going on about, the doctor wonders.

This imaginary and yet very possible scenario illustrates the importance of language in medical encounters. Although we do not often stop to think about the roles language plays in the delivery of care, words pervade medical systems. The word 'cancer' itself conjures up images of some malign organism eating away at one's flesh, hardly reassuring to the young woman. As the doctor speaks, clearly dominating the conversation, the specialized jargon of biomedicine over-whelms and confuses her. She recalls gruesome narratives she has heard about others who have had to go through what she now faces. Meaning-laden words surround her and her experience.

Language is, of course, an extremely important part of culture. It is the primary way in which ideas are communicated, how new generations learn what their culture is. Language gives meaning to the world; the problem is, we speak different languages, both because some people speak English and others Chinese, and because speakers of the same language often really do not understand each other. This latter difficulty arises in part because words and phrases can have different meanings when spoken by different people and in different contexts. What the young woman and the doctor understand by the term 'breast cancer' are probably completely different things.

In this chapter we will explore the importance of language to health, especially as it is spoken in places. First, we will show how words can be used to name things and thus give them meaning or 'norm' them. Words provide meaning to medical situations when they are used to classify diseases or treatments or when they are used in healing rituals. Metaphors, the subject of the next section, are special words that resonate with meanings that are drawn from many different aspects of our daily lives; they strongly influence how we interpret experiences of illness and health. Another important use of language is in telling stories about illness: listening carefully to narratives can reveal a great deal about what a person really believes or how a person really feels. The later portion of this chapter takes up the related topics of how knowledge, which is often expressed in language, is gained, how it is used in power relationships, and how the power of language might be resisted. These ideas are illustrated by medical encounters that involve doctors, patients and others, who may be using different language systems or discourses. We also show how language and health come together in particular places. Finally, we discuss how language, health and place are linked in two types of media, the novel and television.

We can begin to see how people use language to give meaning to health and disease by a brief look at *medical semiotics*. This approach originates in a concern that the ancient Greeks had in interpreting signs of bodily dysfunction. '[M]any, if not most of the signs that refer to illness states are found outside the body,' Staiano (1979: 111) states, and goes on to say that 'such signs . . . often perceptively integrate biological signs produced by the body with "social signs" indicating disturbances in the "social field".' That is, semioticians attempt to find analogies between biological or genetic codes and cultural or behavioural codes (Staiano 1981). In this way of looking at health, the report of a symptom is an act of communication, an attempt the sufferer makes to say something about himself or herself. Many people attempt to interpret symptoms, including the patient, the healer or diagnostician, and friends and relatives of the patient. Making decisions about seeking care involves using a grammar that has agreed-upon rules. Biomedicine has its rules for assigning a particular symptom to a particular problem (e.g. chest pain = heart problem), but so do other, alternative systems, leading to a pluralism of possible interpretations and the necessity to negotiate among them.

Words, words, words

What's in a name? A great deal, a geographer would say, especially if one is talking about place names. 'Place names are part of both a symbolic and a material order that provides normality and legitimacy to those who dominate the politics of (place) representation' (Berg and Kearns 1996: 99). Naming is norming; that is, giving a place a name establishes the meaning of that name as an 'understood truth' or what is accepted as normal within a culture. So it is with cultures of health.

Because names have meaning, they can, like sticks and stones, harm people. A study found that staff in three British hospital accident and emergency departments classified patients by calling them either good or bad (Jeffrey 1979). The 'good' label was based mainly on medical characteristics and included those patients upon whom doctors could practise their medical skills. The label 'bad' was based mainly on social characteristics and was pinned on patients who had trivial complaints, took overdoses of drugs, or were drunks and tramps; these people were also called 'dross', 'dregs', or 'rot'. 'Bad' patients were sometimes punished by having to wait for treatment. Hospital patients in another study were classified by staff as conforming, moderately conforming or deviant (Lorber 1975), based largely on how easy they were to manage. Bad patients in this situation were often sedated, sent home, neglected or referred to psychiatry. Some doctors use the abbreviation FLK (funny-looking kid) when they want to indicate that there is something wrong with a child, but do not want to alarm a parent. A name that is used mainly by doctors in residency in US hospitals is 'gomer', which may refer to the *Gomer Pyle* television series or may mean 'Get Out of My Emergency Room' (Leiderman and Grisso 1985). Again, these patients (also referred to as 'crocks' and 'gorks') are difficult to manage and threaten the doctors' competence; they symbolize the gap between a physician's self-image of omnipotence and the reality of the limits of medicine. Note that in all these cases names are used to stigmatize patients and often lead to relatively poor care.

Naming things may have different meanings and therefore effects in different countries. In the USA, the word 'cancer' clearly has frightening connotations, but its former association with almost inevitable death has changed somewhat towards images of possible survival. In northern Italy, however, cancer still has strong associations with death, suffering and helplessness (Gordon 1990). Many people who are tested for cancer do not wish to hear the results because naming the diagnosis is tantamount to a sentence of death; naming, it is believed, brings on the disease or the thing named, and thereby closes the door to hope. Non-disclosure keeps the 'condemned' person within their social world and apart from the death, decay and suffering of the 'other' world. This social 'truth' of the effects of naming may have more power than the biological reality.

Names can be used to commercialize health care. This is seen by the example of the name 'Starship' given to a children's hospital in Auckland, New Zealand (Kearns and Barnett 2000). With its reference to popular television programmes and films such as *Star Trek* (and the good starship *Enterprise*) and *Star Wars*, the name evokes for children familiar adventure stories and being somewhere else, in an alien world. This type of norming can, of course, help to allay children's fears of a hospital and thus be a positive aid in healing. On the other hand, the hospital also uses the name to attract cash gifts from donors. Thus the 'Starship' label signalled 'enterprise' in more than one sense (see the section on metaphors on pp. 77–81 for more on their multiple meanings).

Names are often used to classify, to put the often jumbled items or concepts we encounter into some kind of order. Classifications or taxonomies are useful because they help us to think logically about things, but as Gould (1990: 73) says, 'Categories often exert a tyranny over our perceptions and judgments.' By thinking in terms of separations, we ignore overlaps among categories or distinctions within categories. We saw this in Chapter 4 in a discussion of how difficult it is to separate structure and agency. Taxonomies are *cultural constructions*, theories a culture has about order. Gould (1990) provides an example of his point (see Box 5.1 for two more examples). We divide non-foods into illegal drugs and items we can purchase for pleasure, a categorization that results from accidents of history. This has led to legalizing methadone as a controlled substitute for heroin and outlawing heroin itself, but both are opiates. Also, as C. Everett

---

Box 5.1 Cultures name diseases

Health taxonomies are often culture specific. Take the case of disease classifications. The standard biomedical categories are found in the International Classification of Disease (ICD) codes, which are based on the various biological systems such as the circulatory and nervous system. These codes do change, however, as biomedical knowledge changes (the ICD codes are now in their ninth iteration). There are currently 17 codes: examples include (1) Infectious and Parasitic Diseases (codes 001–139); (2) Neoplasms (codes 140–239); and (3) Endocrine, Nutritional and Metabolic Disease, and Immunity Disorders (codes 240–279) (Hart *et al.* 1989).

The Kamba people of Kenya look at diseases in a very different way, basing their taxonomy on four different systems: (1) system related to ultimate causes, including diseases caused by God, spirits of the departed, witchcraft, oneself, and by someone else (not with witchcraft); (2) system related to treatment, including by specialists and by type of treatment; (3) system related to disease characteristics, including severity, curability, communicability, and heredity; and (4) system related to characteristics of the afflicted, including sex, age and body part affected (Good 1987). Imagine the implications this complex categorization has for disease and its treatment, or what happens when the Kamba are exposed to the biomedical system.

Koop, the former US Surgeon General, has stated, legal nicotine in cigarettes is just as addicting as illegal heroin and cocaine.

Just as names have meaning, so do words spoken in rituals (Tambiah 1968). Rituals are 'a form of repetitive behavior that does not have a direct overt technical effect' (Helman 1994: 224). They express the basic values of a society and communicate these values to others, using various symbols, including words. In the health context, rituals are used to heal sufferers physically, mentally, spiritually and socially. Here are three examples. Katz (1981) finds that ritual is an integral part of the efficient functioning of hospital operating rooms. The elaborate procedures that surgeons and their assistants go through are ostensibly designed to prevent infections, but they also allow autonomy to the participants and help them to deal with uncertain situations. How language is used is just one aspect of operating room rituals. Words such as 'clean' and 'contaminated' take on different meanings at different stages of an operation. At very specific times during a procedure, there is either complete silence, terse commands, or joking and idle chatter; everyone within the operation room culture understands what can be said when (and also where). They can joke about patients and others in the room, but not about the rituals themselves.

The case presentations doctors make on hospital rounds can be viewed as linguistic rituals (Anspach 1988). Doctors use a stylized vocabulary and a standard format when they discuss patients; certain words are used repetitively and patients are typed or normed in specific ways. The language used has a clear hierarchy of value: diagnostic technology is valued most, doctors' observations next, and the patient's account least. During case presentations doctors have a chance to sell themselves and their expertise; at the same time, novice physicians are socialized into the medical system and its ways of speaking and thinking.

Our final example of ritual language comes from a study of the healing practice of a group of Catholic Pentecostals. Csordas (1983) describes healing in this situation as a discourse that involves three tasks that all involve rhetoric or the art of persuasion. First is the *rhetoric of predisposition*, which prepares the sufferer to be profoundly moved when spiritual resources are brought to bear on his or her affliction. Second is the *rhetoric of empowerment*, in which the supplicant is persuaded that he or she is experiencing the effects of divine power. Third is the *rhetoric of transformation*: here the patient is persuaded to change his or her basic cognitive, affective and behavioural patterns. These rituals combine sacred power and conventional psychotherapy and are like *rites of passage*, taking patients on a journey from illness to health. Indeed, for many people illness and its treatment are literally a journey; take the case of pilgrims who travel to healing places such as Lourdes (Gesler 1996).

Metaphors and meaning

A *metaphor* is a figure of speech that uses one meaning system to help explain another one. The first meaning system consists of concrete and well-understood words or phrases, whereas the second is less understood and more elusive. Thus, to express the hardness, wildness and danger of inner-city life, the expression 'concrete jungle' has been used. Metaphors are attempts to make the unfamiliar familiar, to bring meaning and understanding to a chaotic and changing world. Without realizing it for the most part, we use metaphors constantly in our everyday conversations (Lakoff and Johnson 1980). Interestingly, many of these metaphors are spatial (Box 5.2): we often orient concepts by using words that indicate either 'up' or 'down'. For example, health and life are up as in 'She's in top shape'; sickness and death are down as in 'He dropped dead'.

Box 5.2 Metaphors geographers live by

The field of geography abounds with metaphors. Physical geographers anthropomorphize nature when they speak of the 'foot' or the 'shoulder' of a hill, or the 'mouth' of a river. Human geographers refer to 'the heart of the city' and highway 'arteries'. Geographical language has seen a large increase in its use outside of geography. Cultural studies discourse, as an example, is littered with spatial metaphors based on at least these three sets of ideas: (1) location, position and locality; (2) mapping; and (3) colonization/decolonization (Smith and Katz 1993). Anne Buttimer (1982) points out that there are root metaphors which are the basis for the views of the world that Westerners have held throughout their history. These metaphors generate analytical categories and are used when invoking claims to truth. A common view, for example, is *mechanistic* and uses the machine as its root metaphor. This metaphor lies behind the fascination for many geographers with 'spatial systems'. The mechanistic view lends itself to rigorous quantitative analyses and claims to truth based on establishing causal links between variables. In contrast, there is the *contextualist* view of the world, which 'sees the world as an "arena" of unique events, and tries to unravel the textures and strands of processes operative in, or associated with, particular events' (Buttimer 1982: 91). Cultural and historical geographers are perhaps the most likely to think in terms of the contextual worldview and use the arena metaphor; this book is an example.

Trevor Barnes (1992) sees metaphors as interpretive schemes and asks a crucial question: Do they really mirror reality or do they create in part the reality they seek to interpret? The point here is that metaphors are

---

constraining and limited; there are 'facts' that make up reality, but the facts themselves are not as important as how we interpret them, using certain metaphors rather than others. Thus facts are *theory laden*, and meaning is made of them by *interpretive communities* or groups of people who share systems of metaphors. Barnes's illustration of these ideas is a detailed examination of the physical and biological metaphors that underlie two major strands in economic geography, neoclassical economics and Marxism. Another excellent example is how nature has been reinvented over the centuries by Europeans through changing root metaphors: during the Renaissance the earth was likened to the human body, in the Romantic period healing metaphors were used, and in Victorian times nature was 'red in tooth and claw' (referring to the evolutionary idea of the survival of the fittest) (Gold 1985).

How do metaphors work in the health field? Recall from the discussion of medical semiotics on p. 73 that signs for symptoms connect the biological with the cultural or social. So it is with metaphors, as the following example shows (Turner 1975). The Ndembu of Zaire call a tree with a red, sticky gum *mukula*; this word is related in meaning to 'coagulated blood' and 'maturation' (*kukula*) – in particular, women at the onset of menses. When young men are circumcised (note links, again, to bleeding, coagulation and maturation), they sit on logs of mukula wood. Words like *mukula* and *kukula* are metaphors for the abstract ideas of healing and togetherness in personal, physical and social senses. Similarly, in New Zealand the Maori word *whenua* is used for both land and afterbirth. The meanings converge in the ritual burial of placenta in the land and the transformation of this into a sacred place (Durie 1994). Language thus interconnects the body, nature and social relationships.

Our use of metaphors has a profound influence on how we perceive various illnesses. The body, for example, is often portrayed as a machine. Although around half the deaths in the USA are caused by heart disease, this problem is described as a malfunctioning pump, a mechanical analogy that everyone can understand and thus is not very threatening. In contrast, cancer, which is responsible for far fewer deaths, is the great unknown, the 'invisible predator' or an 'alien invader' (Sontag 1978). Arthur Kleinman (1978) tells the story of a woman who was in Massachusetts General Hospital recovering from pulmonary oedema secondary to artherosclerotic cardiovascular disease and chronic congestive heart failure (note the biomedical language!). She began inexplicably to vomit and urinate frequently in her bed. It turns out that she had been told by her medical team that she had 'water on the lungs'. From talking with her husband and father, both plumbers, she had developed a plumbing metaphor for what she believed went on in her body: pipes connected her lungs to mouth and urethra and she could empty her lungs through the pipes. Metaphors are related to treatment as well as disease. As was mentioned on p. 30, a group of older adults in a rural area of North Carolina were asked what they did to treat their arthritis (Arcury *et al.* 1999); some of their answers contained metaphorical language. For example, when people applied ointments and other substances, including turpentine and the lubricant WD40, they used mechanical analogies such as 'oils the joints', 'lubricates', 'penetrates to the bones', and 'penetrates to the joints and limbers 'em up.'

When patients describe their illnesses, it is worth listening to the metaphors they use because these may reveal what lies behind their surface manifestations as symptoms. Kleinman (1988) relates the story of a 57-year-old woman, a painter, who had experienced chronic pain in her upper back and neck for more than eight years (actually, the woman is a synthesis of several patients). The words of this tense, frustrated woman are very revealing:

'You know what I think? The stiff neck is a kind of symbol, an icon of what I need to become: tough, stiff-necked.. . . Have you seen the great Renaissance and medieval paintings of Christ hanging there limp on the cross? Head down, neck under such great stress, arms out.'

(p. 91)

Kleinman takes the image of the crucified Christ and relates it to the woman's family history and childhood and he begins to understand how the severe constraints placed upon her have led to her current chronic pain. Another patient, a police lieutenant, had tried every orthodox and alternative treatment for chronic low back pain, but remained in a state of terror that he was falling apart both physically and mentally. The key metaphor in this case was revealed when the man used language that indicated that he saw himself as 'spineless', unable to cope with his job, his marriage and his childhood experiences. This metaphor does not explain the man's illness entirely because metaphors are limited, but it does help in assessing and perhaps alleviating his condition if he can understand and come to terms with its meanings.

An important reason for the effectiveness of metaphors is their ambiguity, the fact that they can be interpreted in different ways. In other words, they are *multivocal* or give rise to *polysemy* (have multiple meanings) (see Box 5.3 for a case study based on this idea). Laderman (1987) illustrates this idea in her demonstration of how healers among both the Cuna of Panama and the Malays of Malaysia use ritual incantations to aid women in giving birth. Their language contains metaphors relating to various acts of creation – the universe, humanity and babies – that reduce the mother's anxiety, affect her physiologically, and thereby ease the birthing process. Although the woman and the healer may interpret the metaphors differently, they relate to various contexts of the woman's daily life and so their multivocality produces the desired effect.

Box 5.3 Semantic networks

Byron Good (1994) has taken the recognition of multivocality or *heteroglossia* and developed the concept of *semantic networks* as a 'form of synthesis that condenses multiple and often conflicting social and semantic domains to produce "the meaning" of a complaint of an illness' (p. 172). As an example, the US preoccupation with being fat can be linked to many metaphors that carry the meaning of 'lack of self-control' (for example, being overweight indicates that one has 'let go'). Semantic domains are 'deep' because they usually lie outside our cultural awareness, they appear to be natural, they are part of everyday discourse. An extreme result of this cultural phenomenon is anorexia.

Another excellent illustration of semantic networks is Good's (1977) study of 'heart distress' among Turkic-speaking women in the northwestern Iranian town of Maragheh. For the people of Maragheh, heart distress is a disease category (see Box 5.1). Its symptoms are described in physical terms: the heart pounds, it trembles, beats rapidly, feels squeezed, is bored or lonely. It is associated with feelings of anxiety about a variety of problems, including contraception, pregnancy, old age, interpersonal relationships and money worries. In concepts from popular medicine in Iran, the heart is both a central physiological organ (for example, in relation to nutrition and blood circulation) and an organ of emotional functioning. It is the central driving force of the body and feels the stresses of everyday life. These notions have echoes in Western cultures as well, where one may speak of 'heartache' or send a heart-shaped valentine to a loved one.

When women in Maragheh were asked about the causes of heart distress they produced a long list of answers. Good (1977) began to link up the words used through their semantic associations and developed two complex, interlinked domains of experience: (1) female sexuality, potency and pollution; and (2) the oppressions of everyday life. As an example of the first semantic network, heart distress was linked to childbirth miscarriage, which was linked to pregnancy, and so on to blood, dirty blood, pollution, menstruation and uterine blood, contraceptive pills, old age, sorrow and sadness. These associations express

---

as one important nexus of meaning a complex of stresses common to the experience of Iranian women: she is sexually potent and attractive to men; her potency is dangerous and must be secluded; but her fertility and attractiveness are regularly disrupted by states of pollution and ultimately threatened by the coming of old age.

(p. 43)

A proper treatment of 'heart distress' calls not for a biomedical remedy but for a thorough understanding of what a patient's life is like in some detail.

Stories of illness

Narrative accounts, Shorter (1985) finds, are a contextual form of knowing from living within a situation as opposed to knowledge obtained from an external observer. The stories that people tell when they are talking about their illnesses reveal a great deal about what their problems actually mean to them and how their interpretations are culturally based. In Kleinman's (1988: 49) words:

Thus, patients order their experience of illness – what it means to them and to significant others – as personal narratives. The illness narrative is a story the patient tells, and significant others retell, to give coherence to the distinctive events and long-term course of suffering. The plot lines, core metaphors, and rhetorical devices that structure the illness narrative are drawn from cultural and personal models for arranging experiences in meaningful ways and for effectively communicating those meanings.

The story is the individual's personal myth which gives a shape to his or her experience (Box 5.4 illustrates this point with a case study). Although many practitioners are trained to be suspicious of stories, the wise healer listens carefully to them because they constitute the character of illness.

Box 5.4 Arthritis causation beliefs and everyday life

In Chapter 4 we discussed the explanatory models (EMs) a group of rural older adults used in order to talk about their experiences with arthritis. We extend that example here to show how they related their beliefs about what causes arthritis to their everyday experiences (Gesler *et al.* 2000b).

A substantial number of the stories that people told about arthritis causation mentioned a specific job or jobs. These working poor people used their bodies to make a living. They worked hard at strenuous and often repetitive tasks: 'plumbing, working under wet houses, laying on damp ground'; working in a factory where 'sometimes it was hot . . . and then it was real cold'. By connecting their belief that what causes arthritis was cold and damp in their work, people were trying to make sense of or justify all they had gone through. They were saying, in effect, 'I know I am suffering from a painful ailment, but it is worth it because I've made an honest living and I'm proud that I worked hard.'

What is especially interesting about the arthritis stories is that they often express particular perspectives on life that take on a moral tone. The most commonly expressed *morality tale* was the importance of hard work. Many people praised the work ethic and castigated those who were lazy. Another idea was that it was one's moral obligation to keep the family going, to feed, clothe and educate them. 'Yeah, catching cold, many times I have been out, had to go in rain and snow to foot the bills after my husband died . . . to make ends meet,' one woman said. Individual responsibility was another idea that was expressed several times. The moral here was that it was one's own fault if one didn't take care of oneself and the result was arthritis. Also, several people condemned the behaviour of others. One person couldn't understand why a neighbour had arthritis because that person didn't work hard. Here are two statements that take the high moral ground: 'As Mama used to say, the younger generation don't wrap up and their bones and all in the wintertime, they're exposed to the cold weather too much'; and (referring to teenagers) 'If they would learn to dress, just how to dress, and face the facts of life.'

In her account of illness stories told by residents of a marginal and largely mestizo neighbourhood in Quito, Ecuador, Price (1987) shows how narratives perform four functions: (1) they help to solve problems because they transmit technical information; (2) they help listeners to expand and refine their own theories about disease: (3) they focus attention on the caretaker role of the narrator; and (4) they reinforce bonds of mutual support among tellers and listeners. One must listen carefully to illness stories, Price says, in order to pick out what is both included and excluded; what points are elaborated upon; what key words, generalizations and metaphors are used; and what false starts, evaluative statements and hedges are made. Here are just a few examples of what Ecuadorian illness stories reveal about the local culture of health: female family members are expected to bear the main burden of home health care and seeking care outside the home; reciprocity in health care is expected, but giving or receiving too much kindness may be dangerous; clinics are expensive and where only the rich go; hospitals provide inadequate and uncaring treatments; illnesses have three types of immediate cause, physical (e.g. temperature), emotional (e.g. anger) and behavioural (e.g. 'she was playing on the stairs'); and narrators may be most interested in conveying feelings such as 'I did the right thing' in dealing with an illness episode.

Illness stories lie at the heart of the programme of Alcoholics Anonymous (AA) and are mainly responsible for the success it enjoys (Cain 1991). When someone joins the AA they are obliged to listen to and then relate stories that help them to understand how alcohol is controlling their lives. Through stories, they acquire or learn the cultural knowledge of the group and eventually identify with the 'truth' that the group holds, namely that one will always be an alcoholic and that the only way to arrest the disease is to stop drinking completely. They eventually come to reinterpret their life as an AA story. As members of AA, they are obliged to tell their own AA story to others.

Most of the literature on illness stories focuses on what patients have to say about themselves. In a departure from this emphasis, Atkinson (1995) shows how haematologists in a major teaching hospital in the eastern USA tell stories as they discuss evidence about the conditions of their patients. In doctor–doctor interactions, they produce and reproduce knowledge about medical problems: patients are constituted and reconstituted as objects of medical discourse. Like literary creations, physicians' stories may be narrated as mysteries, as atrocity tales, or as moral tales of success or failure. Contrary to what many researchers claim, Atkinson finds that biomedical personnel speak neither with certainty nor with one voice. The voice of science (e.g. a lab technician's report) may conflict with the voice of experience of the older doctor who has seen the condition many times before; the pathologist may disagree with the surgeon. Each tells a different story based on a unique interpretation of the 'facts'.

Knowledge, power, resistance

We saw in Chapters 2 and 4 that cultures of health are based on explanatory models (EMs) of what causes disease and how it can be treated. EMs are, in turn, based on bodies of knowledge that patients and practitioners accumulate about medical matters. It is useful to make four observations about the nature of knowledge. The first is that knowledge is continually being interchanged as people interact with each other: it is *inter-subjective* (Game 1991). In medical settings, doctors and patients in consulting rooms, doctors and nurses in wards, and patients in waiting rooms exchange information about health problems. A second point is that knowledge is *situated* within EMs that are current in specific times and places (Pile and Thrift 1995). Haraway (1991: 42) illustrates this idea when she says that 'It is not an accident of nature that our social and evolutionary knowledge of animals, hominids, and ourselves has been developed in functionalist and capitalist economic terms', an indication that modern scientific beliefs are embedded in current ways of looking at the world. Third, we must engage with the idea which many people hold that there are *undisputed truths*, or that some knowledge cannot be disputed (Fox 1993a). Of course, we would all like to think that we possess the ultimate answers, but claims that a person 'speaks the truth' on a particular matter, as we shall see later, can often be questioned. Recall (Chapter 2) that it is a major postmodern idea that there can be no claims to possess ultimate truth. The fourth point is that knowledge is often *expressed in language*, the focus of this chapter. This idea should make us reflect back on what has been said so far and think of how names, rituals, metaphors and stories (re)produce knowledge.

The possession of knowledge is important because of the simple equation: knowledge = power. One can use knowledge to gain or maintain positions of power or control over others; conversely, those with power can control what knowledge is or is not available (Fox 1993a). As an expression of knowledge, language is also power. In our society, the claim that only rational, scientific knowledge can lay claim to the truth is a dominating idea (Boyne 1990). As Haraway (1991: 43) states, 'Science is about knowledge and power. In our time, natural sciences define the human being's place in nature and history and provide the instruments of domination of the body and the community.'

In no other sphere of life, perhaps, does the power of knowledge (and, more specifically, language) manifest itself more than in health care. Here is Haraway (1991: 203–4) again: 'The power of biomedical language – with its stunning artifacts, images, architectures, social forms, and technologies – for shaping the unequal experience of sickness and death for millions is a social fact deriving from ongoing heterogeneous social processes.' Power relationships are intrinsic to the medical system. Most typically, the provider dominates the patient, but there are also hierarchies of domination among practitioners (e.g. doctors, nurses and technicians) and among medical systems (e.g. biomedicine and alternative systems).

Some would maintain that it is necessary to have imbalances in power between healer and healed, that the authority and expertise of those trained in biomedicine is required for adequate medical practice (Maseide 1991). Others would argue that dominant–subservient relationships are detrimental to patient well-being, however. DiGiacomo (1987) tells the story of being hospitalized for cancer treatment and coming to realize that knowledge about her condition and its possible treatments was the only kind of power available to her. However, her attempts to gain information were thwarted by the hospital staff, who found her curiosity to be inappropriate behaviour: 'I had to fight for every piece of information I got,' she says (p. 323). She came to believe that doctors felt that patients could not handle the truth and shielded them from it.

Knowledge becomes power in medical situations in many ways. The physician's gaze and speech as she examines the patient 'work in delicate harmony to keep the young patient from resisting the physician's intrusions into his space' (Kuipers 1989: 107). Doctors usually ask the questions and control the topics discussed, seeking both to control the situation and to prevent challenges to their status. The physician is established as the active knower and the patient as the passive known. The doctor attempts to subjugate the patient's feelings to his or her rationality and privileged knowledge (Kirmayer 1988), although this attempt may be contested.

To be more specific about the foregoing ideas, consider these two examples of the exercise of medical power. In Anspach's (1988) work on case presentations during ward rounds, she demonstrates how doctors wield their authority by (1) depersonalizing patients (e.g. 'Baby Girl Simpson was the 1,044-gram product of a 27-week gestation'); (2) omitting the agents or medical personnel who made observations or performed procedures (e.g. 'The infant was transferred'); (3) emphasizing technology (e.g. 'The path reported revealed endometrial curettings'); and (4) treating patient accounts as subjective and barely linked to reality. Whereas physicians 'note' or 'observe', patients 'report' or 'claim'. By using the passive voice, doctors mitigate their responsibility; they also attempt to cover and deflect blame for mistakes. The second example reveals a surgeon's power. Following surgery, patients often do not feel well, because of the double shock of the anaesthetic and the surgery itself, so the surgeon faces the difficult task of convincing the patient that the operation was successful; that is, he or she has to maintain credibility and authority. Fox (1993b) shows how this is done by, as examples, using various props such as screens, writing notes that patients cannot see, performing tests, and conferring the gift of a discharge.

In contrast to these negative uses of power is the example of McGuire's (1983) examination of the power of words to bring about healing among alternative healing groups in suburban New Jersey. Ritual words were used to empower and transform people in positive ways by five types of groups: (1) Christian; (2) meditation; (3) traditional metaphysical; (4) occult and eclectic; and (5) manipulative/technique practitioner groups. Included among the empowering words were *glossolalia* (prayer in 'tongues'), mantras, metaphors, and the phrase 'I AM'. Language was used both as a means of communication and as a vehicle for energy or power (e.g. from a supreme being). People believed that ritual words could really affect their everyday life; they helped them to find order, meaning and control.

Knowledge and language are powerful and can be used to dominate, but dominance can also be contested or resisted. Patients can and do react against what medical practitioners say to them. The recent rise in consumerism in health care has tended to temper somewhat the physician's power, for example (see Chapter 8). Who can say that science always 'speaks the truth'? After all, scientific paradigms change and therefore explanations of how the world works change; compare views of the world before and after Einstein. Furthermore, language is often ambiguous and can be interpreted in different ways by people with different EMs. It is seldom the case that an oncologist and a cancer patient mean the same things by the word 'cancer'. This line of thinking takes us back to the postmodern notion of *difference* (see Chapter 2). Even physicians discussing the same evidence about a patient disagree about what it means (Atkinson 1995). Recognizing difference provides the potential for resistance, for challenging someone else's interpretation.

Medical encounters

The role that language plays in health is perhaps most clearly illustrated in medical encounters as patients meet doctors, or as practitioners or patients meet each other and talk. Medical encounters between two or more people are generally viewed as problematic for several reasons, but we shall concentrate on two language-related difficulties: (1) conflicts that arise from different medical beliefs, and (2) differences in power (Helman 1994; Atkinson 1995) – or what Lazarus (1988) calls the *explanatory models* and *critical medical anthropology* approaches, both of which have already been discussed in this chapter.

Recall that the *explanatory model* (EM) approach holds that different actors in a medical situation will use language that often reflects different ideas about what disease or illness is and what should be done about them. When doctors meet patients, rationality, emphasis on physiochemical data, and a view of disease as a single entity encounter subjective meaning given to experience, personal feelings, and disease as embedded in one's lifeworld (Helman 1994). Doctors – bright, highly trained achievers, goal-oriented people with relatively little training in social science or the humanities – converse with lay people, usually less educated, who obtain their information from friends and family, folklore, the media and waiting room pamphlets (Lazarus 1988). As a result, patients and doctors often talk past each other because their language worlds are so different. When doctors talk with doctors, there are EM differences as well: the language of science used by the eager young intern may collide with the more subjective, but also more authoritative, language of the older, experienced physician (Atkinson 1995).

The importance of EMs can clearly be seen if we follow Fisher's (1991) comparison of how a nurse practitioner and a doctor talked to two young women patients in medical consultations. The consultation involving the doctor reinforced family values, technical solutions to problems, and failed to address the patient's fears. The nurse consultation was more open-ended, encouraged resistance, expressed social concerns, and provided a space for the patient to talk. Fisher stresses in this study the roles of nurse practitioner and doctor ideologies and also their relative location within the medical hierarchy.

The *power differential approach* focuses on how knowledge and power are brought to bear unevenly in medical encounters. Because the doctor–patient relationship is asymmetrical, power becomes domination (Lazarus 1988).

> Practitioners, because of their specialized knowledge and technical skill, have information which patients need to reach their decisions. In addition, through the strategic use of language during the medical interviews, practitioners can control patients' access to and understanding of this information.
>
> (Fisher 1982, p. 78)

Waitzkin (1989) accuses doctors of using their position of dominance to support the existing social order: for example, they may encourage men to get back to work and women to stay at home. The uneven encounter that results from this situation may lead to resistance and conflict; recall DiGiacomo's (1987) struggle to gain information about her cancer treatments from a very reluctant hospital staff. Conflict can, of course, be destructive, but it may also result in productive negotiation. In fact, DiGiacomo believed that her oncologist came to enjoy the 'intellectual fencing' with her as she argued her own views on treatment.

The possibility of negotiation leads us to consider other positive outcomes of medical encounters. Suchman and Matthews (1988) see the benefits of contact for meeting the basic human need for meaning. Sympathetic doctors can ease a patient's isolation and despair. They can learn to use explanations that connect patients' personal experience to their cultural models. However, to do this, they must relinquish their control. Tuckett *et al.* (1985) make the point that both patients and doctors have expertise which they need to share. Doctors can make use of their power to induce patients to discuss their problems openly. Both parties to an encounter must realize that the other may be anxious and frustrated.

And what about place?

So far this chapter has discussed the many ways in which language makes an impact on the delivery of health. But where does place fit in? We begin to answer this question by linking language and place and then we put language, place and health together (see also Gesler 1999). Language and place have a reciprocal relationship. Places such as hospital waiting rooms, a doctor's consulting rooms or a native healer's hut give rise to specific types of conversations. Conversely, language helps to create place: naming places helps to make them familiar and also imbues them with power; myths heighten the meaning of place, and language excludes other places (Tuan 1991). To further make this point, Pred (1989) argues for the importance of the use of language that is embedded in situational practice and power relations in places.

Although it is relatively easy to find links between language and health care in the literature, as the many examples cited in this chapter demonstrate, place is often not mentioned at all. If it is referred to, it may not be referred to in a specific way; as examples, Kuipers (1989) speaks of *situated* medical encounters or institutional *settings* in which language is used, and Tuckett *et al.* (1985) say that the setting or situation may be as important to the effect of language as power relations. It seems that these researchers are referring to specific places, but one often has to probe for place in the language and health literature.

One way that place comes through in language and health studies is when a health setting is spoken of as a place apart. Thus DiGiacomo (1987) finds that language use, as well as other unfamiliar practices, makes a hospital a foreign place, another world in which patients have difficulty making their way. Lorber (1975) says that the patient is an outsider in the physician's place of work and thus at a disadvantage. And Mishler (1984) claims that because doctors usually talk to patients in emergency rooms or hospital wards, they have difficulty in entering into their lifeworlds.

Occasionally, one does come across a specific reference to place, as for example when a Catholic group use a 'healing room' for their therapy sessions (Csordas 1983), and illness stories are often anchored in specific times and places (Price 1987). In addition, there is an awareness on the part of some researchers that the uses and meanings of language change from place to place. Cain (1991) found that Alcoholics Anonymous illness stories varied over time and depending on the places where they were told. Atkinson (1995) noted that conversations among doctors varied a great deal as the conversants moved from place to place in a hospital on their daily rounds, and that each place revealed its unique information. Anspach (1988) describes the historical change from physician diagnosis taking place in the patient's home to diagnosis in a hospital or laboratory and how that change turned medical practice from a reliance on the patient's account to a reliance on the doctor's clinical perceptions.

The point was made previously that language could be used as power in medical encounters. Power is also manifested in places. In his discussion of power relations, Foucault says that power should be examined at the sites where it is actually brought to bear; power is localized through its action on individual bodies (Fox 1993a). This idea of the localized wielding of power is echoed in analyses of doctor–patient relationships (Maseide 1991). And the physician may manifest power differently in different places such as a private office, a public clinic, an emergency room or a nursing home (Lazarus 1988). The power of language can also be wielded at the scale of a community (Box 5.5).

Specific places provide an environment or context for medical encounters and thereby affect the language spoken in them. Such physical features as wall colour, arrangement of furniture, or the amounts of light and heat in a room can influence the tenor of a conversation. As Spencer and Blades (1986) contend, buildings, and rooms within buildings, can be designed either to constrain or to enable people. Who is willing to speak openly in a shared hospital room, even with a curtain pulled around the bed (Kirmayer 1988)? What is the effect of sitting in

---

Box 5.5 Ascribing disease to people and place

Susan Craddock's work on disease epidemics in San Francisco during the nineteenth century provides an analysis of the power of language to create images of people and places as diseased that have devastating effects (Craddock 2000). Following a severe epidemic of smallpox, the San Francisco Board of Health formed a committee composed of businessmen and physicians to investigate the area known as Chinatown. The committee's report, issued in 1885, used language that inscribed the district with a pathology that served to further the hatred non-Chinese citizens had for the Chinese and their 'place'. Thus, just as homosexuals, foreigners and the poor are often shown to be deviant, disease too can be used to indicate deviance. Ascribing disease medically legitimates already entrenched prejudices. Pathologizing the Chinese and Chinatown was used as a political tool and a form of discipline to keep the Chinese in their place and their district confined.

Despite the fact that smallpox originated from within the white community and thus outside Chinatown, and that in three out of four epidemics the Chinese death rates were no higher than in other areas, the health department used the committee's diagnosis of the situation to lead several public health forays into the area. Like a doctor treating an infection in the body of a patient, public health officials cleansed, partly tore down, and rebuilt the Chinese community, creating untold suffering in the process. However, disease itself contests the discourses formed around it as it cannot be confined to boundaries; it breaks out and enters other communities, creating 'border anxiety'. Thus, the people of San Francisco had to face the fact that their scapegoating rhetoric falsely limited disease to one ethnic group and one place.

---

a GP's consulting room with its scientific books and journals, computer, landscape paintings and family photographs (Helman 1994)? Or what positive health
effects might conversing with one's friends in a hospital waiting room about a
variety of community concerns have (Kearns 1991)?

A very appealing way to look at the role of place in language and health is as
an arena or stage where a drama is being played out. That is, language is being
used as a performance by various actors in medical encounters. We get a
good sense of this idea when we look at the highly orchestrated silences, terse
commands, jokes and idle chatter of the operating theatre (!) (Katz 1981).
Television of course makes much of medical drama in its popular series such as
ER and *Chicago Hope*. Emerson's (1970) study of what goes on in gynaecological
examinations (which take place in special rooms) is an excellent example of high
drama: she speaks of doctors and patients fumbling for lines, actors producing
scenes, and people coming in conflict and creating dramatic tensions (e.g. how
modestly dressed should the patient be?).

To cap this discussion of the importance of place, we describe Fisher's (1982)
analysis of patient–doctor conversations about whether or not hysterectomies
should be performed. These talks took place in two clinics within the Department
of Reproductive Medicine in a university teaching hospital, the Faculty Clinic
and the Community Clinic. What is of primary importance here is that the two
places had very different contexts or environments and thus the outcomes of
conversations were very different. There was a separation of private and public
space, and examination rooms were separated from consulting rooms in the
Faculty Clinic, but no such divisions were made in the Community Clinic. The
Community Clinic was staffed by residents, the Faculty Clinic by professors of
reproductive oncology. Most patients were referred to the Community Clinic
by social agencies or other clinics within the hospital system; most patients who
came to the Faculty Clinic were referred by other medical practitioners.
Community Clinic patients, typically older, Mexican or Mexican American
women, were more likely to receive radical as opposed to conservative treatment when compared to the generally younger and Caucasian women who
attended the Faculty Clinic.

Health and the media

Much of our knowledge about health matters comes to us through various media:
television, books, pamphlets, and many other similar sources. This is an important aspect of the fact that, increasingly in our postmodern world, less of our
information comes from direct experience and more is *mediated* or interpreted
for us. Mediation can and does bias information and can be deceitful and harmful,
but we can also learn from careful readings of what the media tell us in their

audio and visual messages. We illustrate the role of the media with two examples, one from television programmes and one from a novel.

Medical soaps

It has often been said that television is the most influential medium in many cultures. Indeed, in societies all over the globe, watching television is the predominant use of leisure time. Adams (1992) has argued that television can be seen as a gathering place, a place without a location, where, in a virtual sense, people come together to share cultural experiences. Television thus creates centres of meaning for people. It confers value on persons and objects. Furthermore, culture and television reinforce each other: cultural practices influence what is seen on television, and what TV producers choose to show influences what people think and do.

As well as being a kind of place in itself, television shapes representations of specific places (Mander 1978). For example, Los Angeles was depicted in certain ways (e.g. associating the city with gangs and racial violence or with celebrity status) during endless hours of coverage of the O. J. Simpson trial (Alderman 1997). Often, television attempts to heighten the events of daily life by creating scenes with fast action and intense drama. It is selective, choosing stories to tell that it feels will engage an audience, and at the same time excluding other stories. Critically watching television can be instructive because, like other media, it can reveal cultural norms, societal structures and ideologies (Aitken and Zonn 1994). It is a *semiotic landscape* that can be read for the meanings it conveys. It uses certain discourses and narrative conventions that lend themselves to critical interpretation. It is also the site of struggles over control of what images it will portray.

Some of the most popular television programmes produced in the USA have been 'medical soaps', or dramas depicting the work of physicians and other health professionals in modern hospitals. Shows like *Ben Casey* and *Dr Kildare*, all the rage in the 1960s, reflect the themes outlined in the preceding paragraphs (Turow 1997). Casey and Kildare, not surprisingly, are young, white, male doctors who practise in big city hospitals. The shows' producers developed a formula that combined strong support for biomedicine (a form of medicalization) and popular tastes for melodrama (the very stuff of hours of, usually daytime, soap operas). Hospitals in these shows were depicted as places with high-tech equipment and unlimited resources. Invariably, glamorous male physicians were assisted by equally glamorous female nurses, reinforcing patriarchal stereotypes. Non-hospital settings and non-biomedical personnel, such as those employed in patient care in the home, and alternative practitioners were excluded.

Medical soaps confer value on youth, glamour and high-tech equipment and remove attention from other people and objects that are also important components of health care delivery systems. They both produce and reproduce popular conceptions of what goes on in hospitals. In contrast to the boredom of hospital life for patients and often unpleasant relationships among staff and between staff and patients, medical soaps portray constant high drama and caring attitudes. Contemporary issues are the focus, but situations that make biomedicine look bad are avoided.

The American Medical Association (AMA), representing the biomedical establishment, was heavily involved in the early doctor shows. In the 1950s and 1960s, doctors were becoming increasingly worried by challenges to their power. Their control over patients was waning, hospital administrators curbed their authority, and the federal Medicaid programme further eroded their domination of medicine. Part of the AMA's reaction to these perceived problems was to try to shape the images seen on television screens. The AMA was particularly concerned that relationships between physicians be depicted as non-conflictual and that doctor–patient relationships should display proper respect for the former and deference from the latter. AMA officials set up an all-male, ten-person Advisory Committee for Television, Radio and Motion Pictures which guided shows like *Casey* and *Kildare*.

Medical soaps have risen and declined in popularity over the years. Today, *ER* and *Chicago Hope* (and regional variants such as *Shortland Street* in New Zealand) indicate a revival of the genre. More sophisticated, these shows make a great deal of their attempts to deal with current issues and their attention to technical details. At the same time, they continue the tradition of creating an appealing, if unrealistic, virtual reality that strongly influences the popular conception of how health care is delivered.

Imaginative literature

Humanist geographers have suggested that one can learn a great deal from reading imaginative literature (Pocock 1981). We argue that, apart from examining people's experiences of disease and health in the real world, a great deal is to be gained by reading fictional accounts. Through the use of the same devices we have focused on in this chapter, namely symbols, metaphors and narratives, writers are able to distil the essence of human experience. As illustration we will discuss one of the most famous novels to take disease and death as a central theme, the German writer Thomas Mann's *The Magic Mountain* (Gesler 2000). In this story, which takes place between 1907 and 1914, Hans Castorp, a young man from a bourgeois German family, travels high into the Swiss Alps to visit his cousin Joachim, who is being treated for tuberculosis (TB) in a sanatorium. Instead of staying three weeks as planned, Hans contracts the disease himself and becomes a patient for seven years, until the outbreak of the First World War.

A great deal of the force of Mann's novel derives from the fact that it is centred on a special place, the TB sanatorium (the Berghof) and its surroundings. It is a place apart, hermetically sealed off from the outside world, the scene of an educational experiment in which young Hans follows a path towards knowledge (Weigand 1965; Travers 1992). Some themes associated with ideas presented within this chapter and book can be extracted from the novel. The first concerns the way in which Hans gains knowledge about illness and death, as well as about health and life. It is clear that acquiring this knowledge takes Hans, 'Odysseus in the kingdom of the shades', along a difficult and dangerous journey. The practitioners at the Berghof try to hide from the patients the gruesome consequences of having TB, to mystify what is really going on. They consciously develop an atmosphere of opulence and gaiety, what one might call a 'contrived therapeutic landscape' (Kearns and Barnett 1999). Informed by germ theory and technologies such as the microscope, they seek to medicalize the patients. But the horrors of disease and death break through and Hans is determined to immerse himself in all the ugliness of TB. As a result, he comes face to face with his own mortality and makes a final decision to embrace life and live it to the full. A lesson for us is that we increase our understanding of illness and health immeasurably if we leave our armchairs and desks and immerse ourselves in field studies that directly involve patients.

Ambiguity is a second theme that can be taken from the novel. One of Mann's techniques is to present opposing ideas in such a way that there seems to be truth on both sides of issues. Thus two of Hans's friends constantly argue philosophical points, the two principal doctors at the Berghof have different views about TB, there is a continuing dialogue between order and dissolution, spatial metaphors of up and down come into play, it is not clear whether the climate cures or brings out disease, and disease is shown to be both degrading and ennobling. The result of all these dualisms is an atmosphere of ambiguity, of not knowing where Mann, Hans, or oneself as reader stands. What lessons can be learned from this literary strategy? From a postmodern viewpoint, one can invoke Derrida's notion of *differences*, one aspect of which is that the meanings of words or ideas arise only in opposition to other words and ideas (Cilliers 1998). Furthermore, Mann is trying to show us that seeming opposites cannot be separated: one finds death in life and life in death, order and dissolution coexist. Hans is able to take in and synthesize opposing ideas and arguments, thus providing new knowledge in a dialectical manner.

Perhaps the most explicitly geographic theme in *The Magic Mountain* is that of transitions, or movement from one place or idea to another. As background to the story, Mann describes the tremendous upheavals that the world was experiencing at the turn of the twentieth century in terms of historical events, scientific discoveries, social relationships and economic circumstances. Hans himself experiences an alchemical transformation from ignorance to understanding of disease and health, death and life. Throughout the story, the image of crossing a threshold occurs as a leitmotif: Hans crosses international borders, a patient screams with sudden knowledge of her fate as a priest comes into her room, Hans 'crosses the Rubicon' in his relationship to his beloved Clavdia, Joachim makes the 'short crossing' to death. Hans must leave the immediate environs of the Berghof, against the rules, to 'take stock' of his ideas and make discoveries beyond the range of ordinary human thought. One can argue, as well, that Hans goes beyond the biomedical to examine the true significance of illness and death (Stern 1995). Medical facts, he finds, are often equivocal (e.g. what does the X-ray or thermometer really tell us? To what extent is his disease physiological or psychological?). Delving behind scientific discourses on how the body works, Hans discovers the poetry of living things. In contrast to the Berghof doctors, Mann recommends to the reader no medical therapies (Bauer 1971); rather, he forces us to think about illness and health in new ways.

Listening to language

Words are such a pervasive part of our everyday lives that we seldom stop to listen to them carefully and think about what they might mean. This chapter has put across the idea that language matters because it expresses in many different, often subtle, ways what illness and health really mean to people. And it does this in places. How do we listen to language? For a start, we can simply pay more attention to what various aspects of medical systems – patients, diseases, rooms in hospitals – are called because these words imprint these things with symbolic meaning. One kind of word or phrase to watch out for especially is the metaphor, because metaphors attach meanings derived from everyday things to more abstract concepts. We saw how metaphors pervade how we talk about health and health care. Going a step further, we also saw how entire illness narratives or stories reveal a great deal about what a patient really thinks or feels.

Perhaps the most important message to carry away from this chapter is the idea (a dominant one in poststructuralist thought) that knowledge (often expressed in language) is intimately related to power and that we need to understand how knowledge can be used to impose the will of a person or group over others. Along these lines, we discussed the importance of power relationships among, as examples, doctors and patients, and medical staff. The power expressed through language is easily seen in medical encounters between various actors in health care situations.

Paying close attention to language is one more way to understand how people experience illness in places. The chapter ended with illustrations of this point by showing how language and health have been looked at in specific locales. Examples included both real places such as a doctor's office or a hospital ward and the imaginary places that novelists and producers of medical soaps on television portray. We believe that health geographers need to listen carefully to what is being said about health in all of these kinds of places.

Further reading

Barnes, T. J. and Duncan, J. S. (eds) (1992) *Writing Worlds: Discourse, Text and Metaphor in the Representation of Landscape*, London: Routledge. This book is a good read for those interested in how social theories can be applied to geographic work, but it is especially appropriate to this chapter because of its emphasis on language and discourse.

Craddock, S. (2000) *City of Plagues: Disease, Poverty, and Deviance in San Francisco*, Minneapolis: University of Minnesota Press. This book brings together the three key ideas of the chapter – language, health, place – in a stimulating discussion of how the Chinese and Chinatown were pathologized in nineteenth-century San Francisco.

Fox, N. J. (1993a) *Postmodernism, Sociology and Health*, Buckingham: Open University Press. An important book for understanding how postmodern concepts can illuminate how health care is thought about and delivered.

Kleinman, A. (1988) *The Illness Narratives: Suffering, Healing and the Human Condition*, New York: Basic Books. A fascinating series of studies of how those afflicted with severe pain relate their feelings to a well-known psychologist/anthropologist.

6

CULTURAL DIFFERENCE IN
HEALTH AND PLACE

Introduction

This chapter is about how difference is dealt with in health care delivered in particular places. In one sense, analysis of difference has been an important part of the geography of health care delivery since its inception in the 1960s. Many studies have examined inequalities among groups of people living in different geographic areas in terms of, typically, access to and utilization of different types of care (for many examples, see Joseph and Phillips 1984). These kinds of inequalities or differences are easily mapped and are based upon what may be termed *spatial difference* (Barnes and Gregory 1997).

Health care geographers provided a very useful service to policy-makers and the general public by reminding them of the unfair distribution of and access to medical resources. However, spatial inequality studies tended to treat people as aggregates and thus depersonalize them, as well as mask other important differences. More recently, health geographers have begun to express more concern for an alternative kind of difference, *cultural difference*, which recognizes that individuals and groups of people characterized by ethnicity, gender, sexuality and physical or mental disability have different health beliefs, practices and experiences that must be taken into account in research, as well as in the provision of care (Kearns 1995). It is difference in this sense that is our focus here, especially as cultural differences are revealed in places.

We continue this introduction by first showing how cultural difference is constructed and then how it can be deconstructed; in doing so, we demonstrate how negative impacts on health care can potentially be transformed into more positive impacts. The introduction closes with a look at how difference is manifested geographically. The remainder of the chapter details studies on women's health, the health of different ethnic groups, sexuality and health, and health care for the disabled.

96

A major thrust of social theories such as *postmodernism*, *poststructuralism* and *post-colonialism* has been to show how cultural difference has been constructed (Barnes and Gregory 1997). Societies manufacture differences among people on the basis of sexual orientation, gender, ethnicity and degree of disability; they create an ideological atmosphere in which 'we' (the producers of the systems of difference) achieve an identity that is separate from, and superior in many ways to, 'them'. One way this is done is to construct deviant bodies (Box 6.1). At the same time, dominating groups construct 'the other' (Box 6.2). Relationships among those who dominate and those who are subservient are based on power or control, and some constructions (e.g. white, straight, male) are privileged

---

Box 6.1 The deviant body

In their response to the call for a reformed medical geography (Kearns 1993; Chapter 1), Michael Dorn and Glenda Laws (1994) advocated a focus on the identities that people (oneself and others) attach to the human body, thus creating a politics of difference. This concern echoed a fascination in the 1990s with the body as a surface to be mapped, inscribed upon or marked (Pratt 2000b). Attention was paid to the body as the container of individual identity, separated from others, and yet with a permeable boundary. In his work, Michel Foucault (1975) spoke of the body as an object of power relations; that is, people attempt to control and discipline bodies that have been marked, for example, as black, female or homosexual. What is most important for this text is that medicine constructs deviant bodies, ones that fall outside its definitions of what is healthy. Deviant bodies are constructed by ascriptions of disease. For example, gay male bodies are pathologized as the sites of infection from HIV/AIDS (Craddock 2000). Unfortunately, the focus of medicine on the body shifts attention away from the social conditions that produce disease in the first place.

There are geographies of the body. For a start, the body can be seen as the 'geography closest in' (Nast and Pile 1998). Bodies move into and out of places. Medicine attempts to fix bodies, but they are uncontrollable and indeterminate; like Alice in Wonderland, they shrink and grow, are sometimes in place and sometimes out of place. Furthermore, places become known as diseased because they contain diseased bodies; thus Chinatown in San Francisco became stigmatized as the home of diseased Chinese bodies in the nineteenth century (see Chapter 5).

Box 6.2 The other

Humans have a common tendency to contrast themselves to socially significant others (Jackson 2000). We draw circles or boundaries around ourselves and others we like or agree with and exclude those we see as different or deviant. This happens continually as we carry out our daily lives, as well as at wider geographic scales. Edward Said's publication of *Orientalism* (1978) drew the world's attention to the way Europeans had constructed the category of 'the Orient' to express both their fear of and their fascination with 'the other'. Otherness is used to stigmatize such categories as race, mental illness or physical disability. It involves power relationships: one person or group attempts to establish and maintain dominance over others.

There is a very clear geographic dimension to otherness. Thus Tim Cresswell (1996) shows how people are often preoccupied with their proper place in both the social and the spatial senses, often both at the same time. For example, in the 1980s, New York mayor Ed Koch introduced an anti-loitering law in an attempt to move the homeless (seen by society as undesirables) from public spaces (where normal or good people had a right to be), ignoring the social and economic processes that created homelessness. The notion of being 'in place' expresses ideologies of what is right, just or appropriate by members of a society. In contrast, some people transgress these ideologies and cross social and geographic boundaries, becoming 'out of place'. Resistance to being told what one's proper place is therefore questions that designation.

Some instances of social and spatial exclusion are well known. Notorious examples include the apartheid system in South Africa and the 'final solution' to the Jewish 'problem' in Nazi Germany. David Sibley (1995) points out that there are many exclusions that are less obvious because they do not make the news or are taken for granted within our daily lives. One example is the surveillance of indoor shopping malls by private security firms, looking for teenagers who may not be consuming much and who create fear among those who are there to buy. Sibley also makes the important point that the boundaries that separate us from them are often ambiguous because they are interpreted differently by different people and are subject to alteration. What is your ethnicity if you come from a mixed marriage? What are the boundaries between public and private space where you live?

over others. As examples, males in patriarchal societies dominate females, colonizing European powers subjugate 'uncivilized' natives, gays are forced by homophobic societies to remain in the closet, and disabled persons are disdained, stigmatized or ignored.

The strategy often used in constructing cultural difference is to *naturalize* it or make it appear as though it is only natural, the way of the world, an understood truth, not subject to question. Naturalization serves also to *legitimize* a system of difference. Thus a sovereign defends his suppression by appeal to the divine right of kings, a wife-beater appeals to the Bible for his authority, or a colonial master rationalizes his harsh treatment by a resort to *social Darwinism* (the idea that social inequalities are a result of the biological process of natural selection).

The construction of cultural difference involves a *discourse*, a way of defining categories, speaking about oneself and others, that puts into language the naturalization and legitimization of dominance hierarchies. One sees this, for example, in the metaphors used to express colonization (e.g. invasion, conquest, pacification, possession and control of territory) (Smith and Katz 1993). In the realm of health, studies that examine the geography of AIDS distribution and diffusion reveal a discourse that often focuses on the virus rather than the patient, and thus the gay person may be viewed primarily as a host or a vector to be factored into a disease ecology (Brown 1995; Kearns 1996). AIDS, from this perspective, is not 'our' fault, it is 'theirs'; it is others who exhibit high-risk behaviours, not us.

It should be obvious that the construction of cultural difference may have serious consequences for health care, particularly for the person subjugated, the receiver of the inferior appellation. Numerous studies, for example, indicate that in many societies girls receive less health care than boys (McCormack 1988). Colonists are famous for treating their own and native elites first and the rest of the indigenous population second, if at all (Ityavyar 1987). Because health practitioners are often afraid of the stigma attached to homosexuality, gay men with AIDS often receive inadequate treatment. Disabled people have to fight apathy and distaste for infirmity to claim their share of care (Dyck 1998). As we saw in Chapter 5, patients labelled as 'gomers' or 'rot' are usually poorly treated by hospital staff.

As we come to understand how difference is constructed, we can also think about how it can be deconstructed, opposed or resisted. The ideologies that set up dominance hierarchies can be revealed and contested. Social movements that demand gay rights, civil rights for minorities and equal treatment for women are all attempts to break the bonds of the tyranny of us and them, or us = better, more powerful and more deserving than them. It becomes possible to denaturalize, delegitimize and make space for the marginalized. Discourses of dominance can also come under attack. The argument can now be made that no one group

99

has sole claim on the truth (truths are *not* self-evident), that there are no meta-narratives (Kearns 1995). Rather, there are many voices that should be listened to, none privileged above the other. The texts of discourses are decoded, multiple texts are examined, and texts 'conspire with other texts to reveal new possibilities' (Fox 1993a: 103). Now, instead of allowing cultural difference to confine and discriminate, one celebrates difference.

The critical approach advocated in this book leads us continually to reflect upon or criticize statements we have made. Thus, it is important to point out that discourses of difference can also be criticized. Deconstructing difference, however, is not unproblematic. For one thing, the tendency has been to speak in terms of dichotomies: woman/man, gay/straight, disabled/whole, black/white. Thinking this way, however, masks differences within categories as well as the possible presence of more than two voices; there are degrees of disability along a continuum, for example. Another difficulty has been the attempt to solve the problem of dominance relationships by reversing them, a practice that falls into the trap of recreating the original difficulty in a new form. Third, there is the problem of the tension between constructing new identities (especially for the formerly subjugated) or 'placing' people on the one hand, and allowing for mobility or the freedom to move among identities on the other. Finally, although 'the other' now becomes the focus of study, rather than oneself, there is the almost intractable problem of how to 'convey otherness without reducing other people to variants of ourselves still positioned at the centre and as the norm' (Barnes and Gregory 1997: 438).

Geographers have taken ideas about the construction of cultural difference and shown how they are manifested spatially. That is, places and spaces become ethnicized (e.g. in urban ghettos), gendered (e.g. 'woman's place is in the home') and sexualized (e.g. gay people are tolerated only in certain neighbourhoods) as others are 'put in their place'. Furthermore, reading geography's past, one finds that it has a history of racism and sexism in teaching and research. Conversely, however, geographers can open up spaces and places for what formerly was 'the other', they can listen attentively to plural voices expressing themselves in local situations. This is certainly the case for health geographers, as the remainder of this chapter intends to show.

A woman's place in health care

Ralph Ellison's famous novel *The Invisible Man* chronicles the neglect of African Americans in US society: no one notices or cares about their plight, they may as well not be there at all. When it comes to health care, one could also write a book about invisible women. In the past, the bulk of health care research has been carried out on men. For example, the well-known Whitehall studies carried out in Britain were based on a sample of 10,000 male civil servants. Within geography, the problem begins with the under-representation of women in geographic writing in general (in human geography textbooks, for example), despite the many contributions women have made to the field over the years (Mayer 1989). Recently, some textbooks have included a chapter on a geography of gender. However, as Susan Elliott argues (1998), what one would really like to see is an entire textbook permeated with gendered perspectives on, say, economic, population and cultural geography issues. The relative lack of interest in women continues in health geography; although, as the examples below demonstrate, this has begun to be remedied in recent years. Work in women's health by geographers represents one of several feminist geographies (Box 6.3).

Box 6.3 Feminist geographies

Feminist geographies arose out of feminism, a political movement that itself was a product of social change and political action (Mitchell 2000). Although one can trace the origins of feminism back at least as far as the 1840s in the USA and somewhat later in Britain (including struggles for women's suffrage and many other basic rights such as property ownership), today's feminism stems more directly from movements such as women's liberation in the 1960s. Very basically, feminism's goal is to overturn inequalities between men and women, to challenge the power relations within social, economic, political and cultural spheres of life that favour men over women (Blunt and Wills 2000). Feminist geographers are particularly concerned about how space is gendered and the ways in which gender is spatial. As examples, the spaces of corporate boardrooms and typing pools are differently gendered, as are spaces in cities during the day and late at night. Feminist geographers also seek to redress the imbalance of women in their discipline and how geographic knowledge itself is gendered.

We speak of feminist geographies in the plural because feminist geographers have pioneered thinking about gender and space in many different ways. Only a few examples can be given here. One direction has been to resist the tendency to think of private, domestic spaces such as the home as feminine and the public arena as masculine, by contesting these associations as well as the very separation of public and private space. Another issue is the *spatial entrapment* of highly educated and skilled women in certain parts of cities that puts them into a position of being tapped as a cheap labour supply (Hanson and Pratt 1995). Feminist geographers have also looked at how popular culture shapes what it means to be male or female:

---

these meanings are clearly capitalized on by firms that advertise everything from cosmetics to weight-loss programmes.

Alison Blunt and Jane Wills (2000) have usefully categorized feminism and feminist work in geography as it arose from liberal, socialist and post-structuralist perspectives. Liberal feminists such as Mary Wollstonecraft (1759–1797), who wrote *A Vindication of the Rights of Women* in 1792, and Betty Friedan, who published *The Feminine Mystique* in 1963, challenged the roles that women were assigned in society and helped to pave the way for women's rights in many spheres of life. Socialist feminists argue that, under capitalism, gender and class oppressions are clearly associated. Capitalism is shown to foster patriarchy or the assumption that men are superior to women and thus should have authority over them. For example, women are assigned lower-paid jobs in the division of labour. Feminist geographers influenced by poststructuralist theories are concerned about how knowledge about gender relations is produced, how gender roles are performed by individuals, how women's bodies are marked by language or discourse, and how gender is inextricably mixed up with other human characteristics such as race or ethnicity, class, and sexuality. Thus feminist geographies are closely connected with many of the other key themes of this chapter.

Another problem that arises in studies of women's health is that, again until quite recently, the focus has been on women of child-bearing age and their reproductive processes (Lewis and Kieffer 1994). In contrast, non-reproductive aspects of health are often made invisible. The concentration on safe *motherhood* can be explained by the central role assigned to women in many societies of giving birth to and raising children. Lewis and Kieffer call for an end to this narrow view of the health of women and its replacement by a safe *womanhood* agenda.

Now that gender differences in health and health care are being looked at more closely, some of the 'conventional wisdom' is being debunked. Elliott (1998) shows that three myths about women's health cannot be supported by the evidence. The first is the old saying that 'women get sick and men die'. An analysis of mortality and morbidity data from the Scottish Twenty 07 survey and the UK Health and Lifestyles Survey showed that, contrary to the saying, gender differences were different for different conditions as well as stages in the life cycle (Macintyre *et al.* 1996). A second myth is that all members of a household have equal access to resources. In many places, however, this is simply not true; that is, women have restricted access to resources and little or no say about how money is spent for such things as health care. The third myth is that women who take on multiple roles (e.g. as worker, mother, partner and household manager) will experience more stresses than males and thereby be at greater risk for health problems. The evidence for this *multiple burden* hypothesis is equivocal, however. One has to look closely at how women *respond* to stresses as well as the stresses themselves.

In keeping with the title of this book, we emphasize in this section that gendered health is strongly related to the cultures of health that exist in specific places; a few examples will illustrate this idea. In rural Egypt, lower-status individuals (often young women) and higher-status individuals (males of all ages and mothers with sons) receive differential health care. Whereas the latter are taken to biomedical practitioners, often distant from the village, the former have to make do with home remedies (Lane and Meleis 1991). In Africa, men and women have about the same incidence of HIV, but almost all the hospital beds allotted to AIDS patients are occupied by men (DeBruyn 1992). McCormack (1988) makes the case, using several examples as illustration, that if females have low status within a society, they will receive relatively little social investment and this will be reflected in higher mortality rates for girls than boys.

An analysis of the issue of where a woman will deliver her child and what practitioners will be involved reveals several of the problems women face in obtaining health care. A good illustration of this is the struggle that took place in the USA over the past century between women, midwives and physicians (Schneider 1998). A hundred years ago, most women gave birth at home with a midwife in attendance. Then midwifery declined as physicians portrayed midwives as dirty, backward and poorly trained. At the same time, the medical establishment was able to put across the idea that pregnancy and childbirth were 'diseases' that had to be managed by experts. It was true that childbirth became safer and less painful for most women, but in the process it became medicalized and under the control of physicians.

In the 1960s and 1970s, as the women's movement encouraged women to take control of their bodies, the assumption of the necessity of medical management of childbirth was questioned. Some women began to seek alternatives, including home births and midwife-attended births. The medical establishment reacted by co-opting these alternatives. More physicians were trained as delivery specialists and hospitals offered more maternity services. However, the home birth movement remained alive, and free-standing birthing centres with certified nurse-midwives (CNMs) were set up in the late 1970s. In turn, some hospitals hired CNMs and attempted either to buy out free-standing centres or to force them out of competition by establishing their own. Midwives are now very much a part of the child delivery system. However, most must agree to be under the control of a hospital, birthing centre, health department or physician group in order to practise. Midwifery has become mainstream biomedicine 'co-opted by the medical establishment as a marketing strategy and as a means of controlling provider costs' (Schneider 1998: 127).

In New Zealand, Abel and Kearns (1991) took up the issue of women's reproductive choices by focusing on a very small (around 1 per cent) but increasing demand in that country for home births. As was the case in the USA, New Zealand witnessed a shift over the twentieth century from a situation where most births were at home to the current almost total control of childbirth by physicians and hospitals. However, legislation reinstated the autonomy of midwives, and a range of alternatives to hospital-based delivery, with home births at one end of a continuum, have been developed. Women's choice of home births comes in response to the increasing medicalization of delivery and a desire by mothers to bond with the child after birth, which is difficult in a hospital setting.

What is it about the home itself that makes it a place where a woman would choose to have her child? This is the research question Abel and Kearns (1991) attempted to answer by conducting informal, in-depth interviews with eight middle-class women in Auckland who had experienced a planned home birth between ten months and five years before the interview. When five of the eight women requested home births, three were actively discouraged by their general practitioner and two encountered reluctant acceptance. It seems that a prospective mother 'must be very assertive, articulate, and in good health to obtain a home birth' (p. 830). Poor women usually do not meet these criteria. Asked why they sought home birth, the women in the study focused on two themes, control and continuity. They contrasted the control over the birthing process they had at home with the powerless positions they would have been in within a medical institution. Also, at home they would be assured of obtaining antenatal, labour and postnatal care from the same doctor and midwife; in a hospital they could never be sure who would be attending them at any one phase of childbirth. At the same time, the home provided continuity of place; that is, the birthing process blended with other daily home activities. Home was experienced as positive and empowering. The moral of these New Zealand and US stories about reproductive health, as well as the health concerns of women in general, is that women must be empowered to make choices about their care and to carry those choices out.

'Empowerment' is a buzzword that has often accompanied the struggle of women (and other subservient groups). The idea is to gain political, economic and social autonomy for the underdog and thereby improve conditions such as health (Asthana 1996). Autonomy, it has been theorized, should come from the grass-roots level rather than from the top down; it should occur in communities or places. Women should be mobilized to create healthier conditions for themselves by educating and organizing them. In practice, however, empowerment may be very difficult to achieve.

Sheena Asthana (1996) also suggests that for women to take collective action requires a sense of common interests or needs. In her study of Visakhapatnam, a city on the east coast of India known for its commitment to gender-aware planning, she found that there was, unfortunately, little evidence of a collective gendered consciousness. What seemed to matter most was culture, which was very place specific. For example, women's low status in relation to men, marriage and motherhood as expected behaviour, and lack of employment opportunities forced women into roles that both undermined their health status and severely constrained their ability to mobilize. More broadly, the social, economic and cultural context of communities within Visakhapatnam, including the history of the place, personal characteristics of community leaders, and the involvement of outside agencies, created an environment that worked against empowerment. However, it was possible for a sense of identity and solidarity to be forged through the development of a sense of place. Despite this, Asthana contends that it might be better for women to promote less vague, less idealistic and therefore more attainable goals than 'empowerment'.

What should an agenda for studies of women's health look like? Lewis and Kieffer (1994) and Elliott (1998) make some suggestions. For a start, more data need to be collected on the health problems and health care of women, in both large- and small-scale studies. Second, women's health should be studied within theoretical frameworks that are informed by social theories used in feminist geography and health geography. The theoretical background should stress the relations of power between the genders that so crucially affect exposure to health risk and access to resources. A third emphasis should be on examining the health of women in the places within which they carry out their daily lives. This would lead to asking the following kinds of questions: How do women's temporal and spatial spheres of activity affect their health? What pathogens or toxic substances is an African woman exposed to when she works in an agricultural field? What are the health risks to an Asian woman who is a petty trader in the informal economic sector? How does a Latin American village woman cope when her husband migrates to the city for work? What environmental risks do factory workers face worldwide? What stresses do women in industrialized countries encounter in the workplace and how do they respond to them? How does a disabled woman carry out activities within her home?

Ethnic differences in health experience

Health care studies often differentiate populations on the basis of *race* (e.g. white versus non-white or European American versus African American), partly because census and other types of data are available for these categories. Here, however, we prefer to distinguish groups of people based on *ethnicity* rather than race

(Box 6.4). It is important to note that ethnic groups are often identified with particular places, as is the case with Southern rural African Americans. Groups with a common culture often come to occupy a fairly well-defined space (e.g. an ethnic neighbourhood in a city), perhaps because they have been forced into this space by racist, discriminatory practices, but also because they wish to be in close proximity in order to share a cultural heritage. At the same time, different ethnic groups may share the same space, in some cases remaining culturally distinct, but in other cases experiencing a great deal of cultural convergence.

Box 6.4 Race and ethnicity

Race is based on biological differences among people. Many anthropologists and other social scientists no longer think in terms of race because there is little scientific evidence for consistent biological differences. However, *racism*, or the idea that members of one race are inferior to those of another (however race may be defined), remains a problem. Racism is a social construction that often becomes institutionalized and leads to inequalities in such areas as health care provision.

Ethnicity is a difficult concept to define, but we use it here because it has a closer identification than race with a homogeneous culture. There are two opposing views of how an ethnic group originates (Isajiw 1974). The *primordial* view is that ethnicity can be traced far back in history, arises out of kinship, and is ascribed to a person at birth. The *instrumental* view holds that ethnicity is constructed by current and local situations and that a person can either acquire or throw off an ethnic label, depending on the circumstances. It appears that there is merit to both these views because each is more or less important for defining specific ethnic groups. For example, both the Hutu and the Tutsi in Central Africa can trace an ancestry back for centuries, the former group to a black African agriculturally based society and the latter group to a Hamitic pastoralist society. Today, after centuries of intermarriage and living and working together, they have a common culture (e.g. in terms of language), and who is labelled Tutsi or Hutu at any one time in any one place is largely a social and political construction. Indeed, the recent instances of genocide or 'ethnic cleansing' in Rwanda were mainly instigated by rival political factions. In the remainder of this section, although we may talk about what are often thought of as races, we are treating the various groups mentioned as ethnic groups with a relatively homogeneous culture (e.g. African Americans in the rural US South).

Ethnicity (and race) is often a conveniently designated 'population factor' that masks the effects of underlying causes of social problems such as poverty which result from an unfair economic system. Examination of the causes of infant mortality, a good indicator of the overall health of a population, is a case in point (Gesler *et al.* 1997). It is easily seen from the data that African Americans and other ethnic minorities such as Latinos have higher infant mortality rates than European Americans in many places. However, one needs to look beyond these simple associations to try to uncover the processes at work. One then finds, for example, that disadvantaged ethnic groups may receive low wages or experience high levels of unemployment, which leads to overcrowded housing conditions and poor sanitation, which lead in turn to respiratory, gastrointestinal and other diseases (Poland 1984). Residential segregation, a result of racist exclusion of 'the other', has also been connected to infant mortality. In a study of 176 US cities LaVeist (1989) showed that residential segregation adversely affected African American infant mortality rates whereas white rates were only slightly affected. LaVeist (1990) also linked poverty (more likely to be an affliction of ethnic minorities) to infant mortality through its effects on nutritional levels for mothers and children, which led to prematurity and low birthweights and thus higher risk for death in the first year of life.

One area in which one can look for ethnic differences is health beliefs. A good example is the effect of religion on health and health behaviour among older adults. Several studies have shown that religious beliefs and practices are related to better mental and physical health as well as better adaptation to illness and disability (Koenig 1994, 1999; Levin 1994). As specific examples, a study in Marin County, California, found that older residents who attended religious services had lower mortality rates than those who did not (Oman and Read 1998), and Ai *et al.* (1998) showed that private prayer appeared significantly to decrease depression and general distress a year after cardiac surgery.

An interesting question that arises from religion and health studies is whether there are cross-cultural differences within the same place (Gesler *et al.* 2000a). Three studies, carried out mostly in rural areas of the USA, give contrasting answers to this question. One might expect that rural US culture would be homogeneous and thus cut across ethnic differences. However, there may be ethnic-specific cultures within the same rural place. A study carried out in eastern North Carolina on religiosity and health among 2,113 European Americans and 933 African Americans found that African Americans and females were more likely than European Americans and males to profess religious beliefs and participate in church-related activities, and that the effect of health problems on both religious participation and mental health was stronger for European Americans than African Americans (Mitchell and Weatherly 2000). In rural Oklahoma, researchers collected in-depth interviews with fifteen African Americans and thirteen European Americans in nearby all-black towns and white settlements (McAuley *et al.* 2000). Here it was found that religion permeated the lives of African Americans far more than was the case for European Americans. In terms of health, some African American older adults mentioned that sin could cause illness, but no European Americans did; African Americans were also more likely to talk about the power of divine revelations to heal.

A study carried out in central North Carolina among 145 European American and African American older adults concluded, in contrast to the Oklahoma and eastern North Carolina studies, that there were no significant differences in the ethnic groups in the use of religion for health self-management (Arcury *et al.* 2000). Members of both groups in this part of the rural South are mostly conservative Protestants whose religion strongly permeates their lives; cultures converge here, at least in some respects. Thus it seems that ethnicity and ethnic cultural beliefs related to health matter more in some places than others. Part of the reason for the seemingly contradictory findings, of course, is that the studies measured different aspects of the religion and health link. However, it is important to investigate under what circumstances there is ethnic cultural convergence within a place and when there is not.

Ethnic convergence or divergence in health beliefs within a place may depend on the specific belief in question. Osteoarthritis (OA) is a chronic, painful and debilitating disease. Very little is known about its real causes, but it probably has a diverse aetiology, and certain risk factors such as age, heredity, accident or injury, and being overweight are well known. Beliefs about the causes of OA have usually been placed into two categories: (1) folk or traditional, with a focus on exposure to the environment (mainly to cold and damp); and (2) biomedical, which includes the risk factors mentioned above.

In a rural North Carolina county, Gesler *et al.* (2000b) studied a group of 136 African American and European American males and females who were diagnosed with OA (see also Box 5.4). They analysed responses to questions that asked what respondents believed to be the cause of OA. About two-thirds held folk beliefs, but African Americans were far more likely than European Americans to express ideas about exposure to the environment. In contrast, European Americans were more likely to mention injuries or accidents and heredity. There were no ethnic differences in expressions about the importance of age. Also, both groups told stories about their experiences of OA in relation to the work they had done throughout their lives in similar proportions. Both groups talked about both biomedical and folk beliefs, indicating a convergence between two belief systems or group explanatory models. Thus we see that within the same county there are similar levels of response with respect to some aspects of OA aetiology beliefs, but different levels of response in other respects.

Geographies of sexual orientation

AIDS is the disease that, more than any other health problem, revealed society's attitudes towards homosexuality (see Box 6.5 for more on the ways geographers look at sexuality in general). When the AIDS crisis became fully acknowledged in the 1980s, medical geographers hastened to examine this usually fatal disease. The graphs and maps that they and other researchers produced helped to reveal the global dimensions of the epidemic and, in a very general way, how the disease diffused. As Michael Brown (1995) points out, however, these geographies, based on standard techniques of spatial science, served to misrepresent, erase or obfuscate much of what is essential to understanding and thereby preventing or alleviating the effects of the disease. The statistics presented to the public made it appear that AIDS was almost exclusively a gay disease, heightening many people's fears about this segment of society. Gay men were thus 'closeted' by public reactions to the spread of AIDS.

Box 6.5 Sexuality and geography

Sexual geographies are informed by concepts developed about sexuality (e.g. by psychologists and psychoanalysts) (Pratt 2000c). The core of the politics of sexuality is to challenge the societal norm of heterosexuality (female–male or opposite-gender sex) and to show how heterosexuality is made to appear natural (*heteronormativity*) and homosexuality (same-sex desire) is made to appear deviant (Blunt and Wills 2000). In Western society, homosexuality has often been legislated against; lesbian and gays have been discriminated against and treated violently by homophobic individuals. In the past few decades, however, homosexuals have made some progress in 'coming out of the closet', or revealing their sexuality, and gaining some rights.

   Sexuality has led to very serious and often violent 'culture wars' (Mitchell 2000). On one side of the debate are those who believe that homosexuality is 'an abomination to the Lord' and should be censured or banned. In opposition are those who argue that people of any sexual orientation should have the same rights. Sigmund Freud (1856–1939) and Jacques Lacan (1901–1981), both psychoanalysts, argued that sexuality, rather than being biologically determined, was a societal construct. Michel Foucault (1978) declared that sexuality was constructed by the discourses that claimed to be describing and analysing it and that sexual designations were used to exercise power over individuals as well as over the knowledge generated about sexuality (Blunt and Wills 2000).

How do geographers study sexuality? One way is to investigate homosexuality, as well as heterosexuality, in both private and public spaces. In general, heterosexual activities are permitted in both types of space, while same-sex activities are restricted to 'private' spaces, if allowed at all. Although lesbians and gay men often seek to be 'invisible', they have emerged in recent years to form distinctive urban spaces; geographers have mapped their residential and commercial neighbourhoods in such places as the Castro area in San Francisco (e.g. see Knopp 1995). However, Mitchell (2000) wonders whether these neighbourhoods are truly liberated or really just ghettos. Another way sexuality has been studied by geographers and others is by contesting the notion of fixed sexual identities. The poststructural argument here is that sexuality is fluid; it can alter over space and time as individuals engage in relationships with others. People behave differently, they perform their sexuality differently, depending on the circumstances of place and time (Blunt and Wills 2000)

Taking standard disease ecology approaches, spatial analysts focused on the virus and how it moved, treating gay bodies as depersonalized biological vectors or hosts for the virus. The treatment of gay men as objects, rather than subjects, Brown (1995) believes, revealed that many scientists were ignorant of and uneasy with the lives lived by this group. After all, homosexuality has historically been stigmatized by the medical profession. Gay bodies came to be seen as threats; ironically, the very people who should have been helped were isolated and condemned. Furthermore, gays were treated (along with intravenous drug users and others) as *high-risk* groups, a way of labelling and stigmatizing people who were already thought to be 'the other' in terms of their behaviour.

What spatial science lacks, Brown (1995) contends, is an understanding of the social construction of AIDS, as well as the social and cultural networks of the places in which gay men carry out their daily activities. Thus he began an eighteen-month ethnographic investigation of local responses to AIDS in Vancouver that revealed geographies of AIDS at the level of the state, the voluntary sector and the family. An important finding was that the gay community had mobilized itself to educate members about safer sex. Most gay men across Canada considerably altered their sexual practices. Rather than acting as agents of diffusion, they began to block the spread of the virus. They used diffusion processes to circulate information about the disease. Ethnographic study also revealed the terrible stresses imposed on gay people and their communities; the stories gay people told about their experiences spoke to their personal dilemmas.

Wilton's (1996) ethnographic study of nine men (eight of them gay) with symptomatic HIV and AIDS in Los Angeles provides further in-depth analysis of the health care problems faced by this marginalized group. Interestingly, although most of these men were 'marked' as being gay, they were also characterized by ethnic diversity: two were Latino, three were African American and four were Anglo. Wilton found that the places where his subjects lived were very important in determining the reactions of others to their condition and what resources were available for their care. HIV/AIDS forced its sufferers to renegotiate their daily life paths and relationships with others. More specifically, their experiences were analysed in terms of a heuristic framework that encompassed the individual, social networks, service providers, the surrounding community and the broader social context.

At the individual level, it was clear that HIV/AIDS imposed severe geographical constraints on the movements of the men. Geographic space was also hemmed in by psychological challenges (e.g. it was difficult for many to go out and meet others because of fear of encountering the stigma attached to them). Study subjects found that social networks, consisting of family and friends, could either be very supportive or have a very negative effect. They found themselves to be very dependent on service providers who demanded much of their time and energy. Although this particular group did not experience much formal opposition from within their communities, some people did reject them and this was very hard to bear; particularly difficult was looking for a new partner when one's disease status became known. In terms of social context, two institutions stood out, government and religion. The men often criticized an uncaring government and some suffered from the religious beliefs of their families which found homosexuality and their illness unacceptable.

Wilton (1996) traced the space–time trajectories of his subjects through five stages following diagnosis with HIV/AIDS. First came the shock of diagnosis, when many feared that they were going to die. Then each respondent began a period of geographic and social withdrawal, the cocoon stage. After some time, life began to assume new meaning as the men became involved in new activities and extended their daily activity spaces outwards once again. Unfortunately, a relapse and a second period of withdrawal followed and health deteriorated. The final stage may be death or permanent hospitalization, but recovery is still possible for some, a possibility that would entail another expansionary period. Thus the time–space path of an HIV/AIDS sufferer is not a smooth one; what is striking is how well the men were able to maintain control over their spaces in the long run.

A third study, again focusing on AIDS, takes us in still another direction. In 1988 Casey House, a unique housing facility for people living with AIDS, opened in Toronto in an inner-city area that people commonly identified with the gay community. Quentin Chiotti and Alun Joseph (1995) set out to 'interpret' the location of this AIDS hospice, using a combination of positivist, structuralist and humanistic perspectives (see Chapter 2) to look at, respectively, proximity, externalities and community conceptions. At the same time, they considered the location of Casey House in terms of space and place.

Spatial or access considerations were that the facility should be close to the populations it was intended to serve, to treatment facilities (usually hospitals), and to volunteer care providers. A Hospice Steering Committee, supported by the Ontario Ministry of Health, had the task of, among other things, finding a good location. They narrowed the search down to an inner-city gay neighbour-hood, thus meeting the access criterion quite well, given the association between AIDS incidence and homosexuality at that time. From a structuralist perspective, Casey House was placed, as expected, in an area where people with little polit-ical or economic power would not or could not object. That is, the neighbour-hood lacked the ability to repel what are generally seen as 'noxious' facilities. A humanistic interpretation of Casey House's location focuses on the meaning it had as a 'place in space'. It is centrally placed within gay culture, and this is significant in the consciousness of local homosexuals. It symbolizes gay resistance to AIDS and empowers a group often marginalized by others.

People with disabilities

Most people, including geographers, carry out their lives and research assuming that everyone is physically and mentally fit. This notion, termed *ableism*, 'refers to ideas, practices, institutions, and social relations that presume able-bodiedness, and by so doing, construct persons with disabilities as marginalized, oppressed, and largely invisible "others" ' (Chouinard 1997: 380). Ableism tends to margin-alize disabled people economically, politically, socially and spatially. It affects their interpersonal relations, their job prospects, the degree to which they engage in political processes, and access to many public and private spaces. Being disabled leads to stigmatization and misunderstanding. Vera Chouinard, who at the time walked with a cane, was once accosted by an older man on a street in downtown Hamilton, Ontario, who shouted, 'Stop using all those damn drugs' (p. 381).

The very word 'disability' indicates a social construction of difference. Groups of people who live in various places at various times decide what a physical or men-tal disability will be. Clearly, then, definitions of impairment or deviance from the norm will change over time and space, as Michel Foucault (1965) has demonstrated for mental illness. At the same time, there may be competing discourses or notions about disability in the same place; for example, physicians and patients might dif-fer over the definition of what constitutes a heart problem or at what stage it becomes a disability. Box 6.6 pursues the idea of definition or labelling further.

People who become known as disabled are particularly vulnerable to unequal power relations. Physicians attempt to impose their own ideas about what they can and cannot do. They may lose a job because a boss is unsympathetic to their

---

Box 6.6 Labelling

At the heart of constructions of disability is the process of labelling: giving a potential health problem a name that has meaning within a society. Labels, as was demonstrated in Chapter 5, are important. For one thing, some labels may carry more emotional weight; compare the rather cruel epithet 'crippled' with the more benign and politically correct 'disabled' or 'impaired'. Some labels, such as 'mentally ill', might attract more attention from medical people, whereas those with 'mental handicaps' might be ignored. Many labels stigmatize or marginalize others, making them feel as though they are undesirable members of society because they are bent over from arthritis or are 'slow learners'. Sometimes a label is more acceptable in one place than in another, as happens when a 'deviant' person is accepted in a hospital setting but not in a public space.

Labelling, or discourses about deviance, can be seen as ways of marking the body or inscribing cultural constructions on the body (Butler 1990; Box 6.1). Furthermore, the way the body is marked may have a great deal to do with a person's identity – how one sees oneself in relation to others. When many people are told they have a disability, their perceptions about themselves change; they begin to worry about how they will be perceived by others, whether they will be shunned, or whether they will be the same active person they once were.

---

plight, or a sexual partner may abandon them because they are no longer perceived as desirable. They may lose a great deal of control over affairs within the home as others restrict their activities and access to resources, perhaps stating that this is 'for their own good'. For many, having a disability affects the control they have over their own bodies.

Within geography, attention began to be paid to disability with a seminal article by Reginald Golledge (1993) and reactions to his ideas. Several years later, Brendan Gleeson's landmark book *Geographies of Disability* (1999) summarized and extended work in the field. Health geographers began to apply ideas from disability to their work; three case studies illustrate this relatively new direction. (These are all micro-scale studies: see Box 6.7 for research carried out over a larger space). In the first study, Isabel Dyck (1998) discusses an investigation she carried out among a group of fifty-four disabled women who had multiple sclerosis (MS). In particular, she examined how these women 'read their home space in light of changes to the corporeality of the body and its representations' (p. 107). People with MS experience profound fatigue and motor and sensory impairments. Because there is no known cure and the course of the illness is unpredictable, they face a great deal of uncertainty about what directions their lives will take.

Box 6.7 Transport and disabled people in Sydney

Rebecca Dobbs (2001) examined the problems disabled people face at the scale of the city. In urban areas the disabled population is very heterogeneous, distinguished by such factors as type of biological impairment, degree to which city environments affect them, and by their location with respect to the places they would like to go to. They face hostile environments on two levels. First, cultural signifiers in the built environment express society's aversion to disability. Second, the structure of the city, heavily influenced by the requirements of capitalism, puts enormous barriers (e.g. heavy emphasis on the automobile, buildings that are difficult getting into or out of) in the way of disabled people. People in wheelchairs and those with such sensory impairments as blindness or hearing loss find getting around the modern city especially difficult.

One of the key issues for the urban disabled person is simply going from here to there. Dobbs (2001) seized on this difficulty for her study of the public transport system in Sydney, Australia. In particular, she wondered how well the train system run by CityRail provided access to the disabled. By law, CityRail and other public services are required to provide access to all potential clients. But has CityRail really done this? To answer this question, twelve stations out of the over 300 in the system were selected for intense scrutiny: four were labelled EasyAccess by CityRail, four were 'wheelchair accessible' but not EasyAccess, and four were not listed as having any level of access. Access was considered on two levels: (1) technical requirements for various needs (e.g. tactile tiles for visually impaired people, ramp gradients for wheelchairs), and (2) a subjective assessment of how 'workable' access was for disabled people. Figures 6.1 and 6.2 give one an idea of some of the specific issues that the study dealt with.

Findings from the study included the fact that there was very great variability in access characteristics from station to station, that EasyAccess stations were not necessarily better than other stations, and that signage (e.g. directions to an elevator) was not always appropriate or easily found. Traffic at several stations was extremely daunting at certain times of day. Staff, who were supposed to help disabled people (e.g. to get on and off trains), were not always available. In sum, access legislation was adhered to only half-heartedly in many instances.

Figure 6.1 An access elevator at Waverton Station, northern Sydney. Local residents agitated for elevators when the commuter train system proposed renovations to the station along historic lines. This represents both a good technical solution to access problems and an unwelcoming signifier, in that the layout and the rock-cut walls combine to create a dungeon-like feeling. Photo by Rebecca Dobbs.

Figure 6.2 Step up from platform to train in Sydney CityRail station. This step is obviously a problem for many disabled people. Only a few (generally newer) stations in Sydney are without this height differential, even though there is no obvious reason for reproducing this disabling design. Photo by Rebecca Dobbs.

Some of the women in Dyck's study had 'hidden' disabilities that were not apparent to outsiders, while others were clearly restricted in terms of their mobility. Their experiences of getting around their homes varied according to the severity of the disease, who else was present in the household, and how they associated their new identities with their home spaces. Because the home presented physical barriers, some women reconfigured spaces within the home or moved to homes that were easier to move around in. Many attempted to continue to carry out their roles as mother or the person in charge of carrying out most household tasks.

A crucial issue for women with MS was the maintenance of their identities as others read their bodies in new ways and they made adjustments to their environments. Many struggled not to be identified among the disabled. Social relationships were altered: in some cases male partners left them, in other cases they showed strong support. Some women were subjected to the power relations imposed by 'experts' on their disease, even though their personal experiences may have run counter to professional interpretations. Although the biomedical verification of MS could help a woman to procure financial and other benefits, some chose to hide their disease from public view; the home could thus become a place of concealment.

Pamela Moss (1997) also chose to study how women with a disability, rheumatoid arthritis, 'contest their home environments through embodied social practices' (p. 23). In her analysis, she came across three ways in which women negotiate home space, exemplified by three women. One woman, Marilyn, actively accommodated her environment to her disability. Fortunately, her level of income enabled her to make modifications such as installing a chair glide up the stairs to the upper floor of her home. She used special devices to grip objects, moved around outside on a scooter, and could afford a gardener. She had a good social network of friends and was active in a charitable organization. As her body degenerated, she accepted its mysteries and made continual adjustments to her surroundings and activities.

Florence made several structural adjustments to her home such as installing grab bars in the bathroom. However, her adjustments were not as successful as Marilyn's. Each day she assessed what she could do and accepted such minor aids as having people at the grocery store loosen jar tops. Although she lived alone, she valued her social networks very highly. The problem was that she engaged in social activities (e.g. eating foods that would later cause her pain) that were detrimental to her health; in other words, she failed to make appropriate social adjustments. Erica, representing the third type of women, felt invisible because of the restrictions imposed on her. Living alone on a small income, she had great difficulty in finding the money to treat her disability and she also discovered that her friends did not understand her difficulties. She put a great deal of energy into activities with others and as a result suffered intense periods of pain. While her problem was invisible to her friends, her deviant body was constantly present to her.

A third disability case study is an analysis of the narratives of intellectually disabled people in Toronto by Glenda Laws and John Radford (1998). As was the case for mental illness, by the 1980s mentally handicapped people were becoming deinstitutionalized and placed in communities in a variety of residential settings. Laws and Radford wanted to know if their study subjects participated in their communities and whether they regarded themselves as active participants. Using a snowball sampling design, they chose respondents from among those whom society had labelled as 'handicapped' and asked them questions about their living situation, employment status, recreational activities and income.

Most of the respondents said that they felt comfortable in their neighbourhoods. Almost all of those who were working reported that they fitted in at their jobs, but fewer than 30 per cent socialized with their workmates. Many of these stigmatized people appeared to be invisible to others. The stories that individuals told about their daily lives revealed an unusually drab existence. Many of them felt that they had little control over their life experiences; it was generally up to others to make friends with them or not. Thus their social networks were very restricted and they were often frustrated because they could find little of interest to do in their free time. In short, they are marginalized, having little interaction with the so-called normal world. They are bound by, rather than liberated within, the places in which they live (see also Hall and Kearns 2001)

Conclusion

This chapter has made the point that there are groups in society — women, ethnic minorities, homosexuals and the disabled — who may be discriminated against by health care providers, as well as by society in general. Of course, there are many other groups that might also be slighted, such as elderly people or poor people. We have to be careful that these groups are not thought of as completely homogeneous in their needs and treatment, however. Such thinking risks labelling and stigmatization on the one hand and failure to consider individual differences on the other.

What has been discussed in this chapter represents some relatively new departures in thinking and research for medical geographers. But it is high time that we and our colleagues look at the kinds of poststructural issues that those in other disciplines are already investigating. Several concepts should guide our thinking. We need to consider how medicine and society focus on the body, the deviant body in particular. We need to be aware of how 'the other' is stigmatized and marginalized and how bodies and 'the other' are labelled. Work on difference should be informed by ideas from feminism, sexuality theory and identity politics. Knowing about these issues can assist health geographers as they go about trying to give the marginalized a voice in obtaining the health care they deserve.

We hope this chapter has raised a constant awareness of 'the other' (in terms of both individuals and groups). We have succeeded in this task if readers can begin to place themselves in the shoes of those with less access to care and less scope to influence the dominant construction of health itself. At the same time, in the spirit of postmodernism, we should be aware that there are no universal health care solutions, simply because difference is a daily fact of life.

Further reading

Barnes, T. and Gregory, D. (eds) (1997) *Reading Human Geography: The Poetics and Politics of Inquiry*, New York: John Wiley. This book contains a collection of readings on such topics as textuality and difference, with very useful introductions to each topic.

Blunt, A. and Wills, J. (2000) *Dissident Geographies: An Introduction to Radical Ideas and Practice*, Harlow: Prentice-Hall. This text introduces the reader to the ideas and practices of geographies of anarchism, class, gender, sexuality and post-colonialism.

Cresswell, T. (1996) *In Place/Out of Place: Geography, Ideology, and Transgression*, Minneapolis: University of Minnesota Press. This book discusses very important ideas about how society attempts to dictate what and where one's proper place is and how such attempts are resisted.

Gleeson, B. (1999) *Geographies of Disability*, London: Routledge. Gleeson's book contains sections on a socio-spatial model of disability, and both historical and contemporary geographies of disability.

Sibley, D. (1995) *Geographies of Exclusion*, London: Routledge. The special value of this book is that it shows how 'the other' is created and spatialized in everyday life through power relations.

7

LANDSCAPES OF HEALING

Introduction

The word 'landscape' is often used in everyday speech – we speak of natural landscapes, of course, but also about landscaped gardens, political landscapes, mental landscapes, epidemiological landscapes (Pavlovsky 1966), landscapes of fear (Tuan 1979) and landscapes of despair (Dear and Wolch 1987). Landscape may appear to be simply 'what is out there' for us to register with all our senses, but its simplicity is belied by the fact that no two people will 'read' or interpret what is out there in the same way. What we say about landscapes is mediated by how we have been acculturated to see the world, what our society believes is important, and by our individual experiences. Cultural geographers have reached no clear consensus on how to define landscape and, in fact, have tried to make sense of the term in several different ways (Lewis 1979; Meinig 1979; Palka 1995a). Here we will attempt to make a virtue out of the plurality of definitions and look at landscapes from three perspectives (see Chapter 2 for more on these approaches), each of which can be illustrated by examples of landscapes of health.

The first perspective on landscapes focuses on the human–environment interaction that lies at the core of cultural ecology. What is studied here is how the physical environment affects human actions and how, in a reciprocal relationship, humans mould their environments to suit their needs. The emphasis has largely been on material aspects of culture, those objects that are visible on the cultural landscape. In this chapter we will discuss the role of nature and objects in nature that are thought to have healing powers. A second perspective holds that cultural landscapes are social constructs that are products of the institutions which societies establish. This approach is based on structuralist notions of the importance of underlying social forces that shape human activities in particular times and places. One particular construct that is currently affecting health care landscapes is consumerism, a phenomenon closely linked to late capitalism, which we will illustrate in the section on selling health as a commodity (p. 125). The third perspective focuses attention on the role of individual human perceptions of what is important in cultural landscapes; that is, landscape is viewed as a personal, mental construct. This approach is informed by humanistic 'philosophies of meaning'. Within this perspective, the meanings given to symbols become an important way to interpret landscapes. Here, we will discuss the concept of symbolic healing landscapes. Following coverage of the three aspects of landscapes there is a section on therapeutic landscapes that takes an even broader or more eclectic view of what makes a place a healing place, using examples of the Navajo healing experience and modern hospital environments.

Nature as healer

It appears to be a common theme within many societies that the physical environment affords healing powers. For many people, simply getting out into nature has a therapeutic effect, whether this entails camping and hiking in the woods or an ecotherapy weekend for stressed-out executives. Western culture is strongly affected by notions of a healing Mother Nature, exemplified by English Romantic poets such as William Wordsworth or John Keats. There is also a persistent *teleological* view of the world; that is, that God will provide for human welfare, and this includes benefits from objects in nature. In the Middle Ages in Europe it was believed that plants similar in shape to human organs could be used to cure those parts of the body. One example of the medieval *doctrine of signatures* was the mandrake, whose shape resembled the entire human form (Sack 1976; Mills 1982). The leaves, roots and bark of medicinal plants are the basis of the pharmacopoeia of both traditional and modern medical systems. It has been estimated by the World Health Organization that 75 to 90 per cent of the medical problems of rural populations worldwide are treated by herbalists (Ayensu 1981).

Part of the lure of retreating into nature is a feeling of 'getting away from it all', removing oneself from the stresses of everyday life. Thus the remoteness of natural settings appeals. When asked where they would most like to recuperate from an illness, many people mention a cabin in the woods or a lonely beach on a tropical island. An ideology that promotes the healthfulness of rural areas as opposed to harmful city life has been a common theme in a rapidly urbanizing world (Box 7.1), although, statistically, rural populations are often disadvantaged in terms of morbidity and mortality rates (Cordes 1988).

Although there is a sense of an overall goodness in nature, we have also seen that specific elements of nature such as medicinal plants and pure air are important. Water is almost certainly believed to be the most essential element that contributes to good health. Water, of course, is absolutely necessary to physical well-being. Perhaps because of this, it has taken on a sacredness that is found

---

Box 7.1 Locating nineteenth-century asylums

Chris Philo (1987) tells the story of the nineteenth-century debate in Britain over where to locate asylums for the mentally ill. The context for this debate was a restructuring of what was called at that time the 'mad-business'. Physicians who belonged to the Association of Medical Officers of Asylums (AMOA), in order to maintain control of provision for persons with mental illness, were attempting to convert a very disorganized system of mental health care into a national system of county and borough asylums. The predominant attitude among planners of the new system was that mad people should be socially and spatially separated from the rest of the population. Although places like Gheel in Belgium, where patients lived with local families, showed that inclusion could be beneficial, exclusion was the rule of the day.

What is relevant to this chapter is the discussion that ensued in nineteenth-century Britain over where to locate asylums. Most planners felt that they should be placed in rural areas, in reaction to the 'dark satanic mills' that the poet William Blake wrote about which dominated the burgeoning urban landscape. The idea was that since those with mental illness were victims of the new industrial society and could not cope with it, they should be taken to peaceful, natural environments. Careful consideration was given to finding a healthy physical environment for patients. Such things as the local climate, underlying rock type, topography, elevation, vegetation cover, drainage and water supply were scrutinized. It was also thought that agricultural work and walks in the fresh air of the countryside would be morally uplifting.

A minority viewpoint held that asylums should be located in urban places. First, this would promote the training of mental health physicians in urban schools, close to where most people lived. Second, poor patients might not be able to travel to rural sites. Third, it would be expensive to provide staff and equipment in the countryside. Finally, urban proponents said that nature could be brought into the city and that urban environments could also be uplifting. Despite these claims, however, rurality and natural environments won the debate.

---

throughout the world. Water purifies, it cleanses the soul, it heals mentally and spiritually as well as physically (Parker 1983). Christians baptize with water, and devout Hindus believe that the heavily contaminated water of the Ganges can heal (Kumar 1983). Places that have achieved a lasting reputation for healing inevitably have water as a central feature. At Lourdes in France, for example, it is the water from a spring that many believe is responsible for miracle cures (Gesler 1996). And, of course, mineral springs are the source of the attraction of spas the world over. In places such as Bath, whether it is scientifically proven or not, many people believed (and doctors encouraged this belief) that specific elements in the waters cured specific diseases such as gout (Gesler 1998).

In addition to water, another of the four elements that the ancient Greeks believed essential to life, earth, is thought by some to contribute to good health. John Hunter (1973) traced the origins and diffusion of geophagy or earth-eating in Africa and the USA. This practice, which existed at least two thousand years ago, may satisfy physiological needs (through ingestion of such minerals as calcium) as well as psychological needs, but it can be harmful as well (e.g. it can cause intestinal blockages). At Esquipulas in Guatemala, the most important Central American Roman Catholic pilgrimage site, Indians mass-produce clay tablets from 'sacred earth' which are blessed by priests at the basilica and sold throughout the entire region to be eaten for physical and spiritual health (Hunter et al. 1989).

A final type of specific objects from nature related to health is animals. Studies have shown that pet ownership has therapeutic value (Tuan 1984). Animals have often taken on powerful symbolic force as aids to healing; a good example is the three animals associated with the ancient Greek healer god, Asclepius (Gesler 1993). The snake was his constant companion and helped him in his dream healings by performing such acts as licking a blind patient's eyes to restore sight. As a legacy, two snakes curled around a staff represent medical practice today. Asclepius also had the aid of a dog, known for its intuitive powers and ability to follow a trail. And then there was the cock, symbol of sunrise, radiance and rebirth, traditionally sacrificed to the healing god. 'A cock to Asclepius' were Socrates' last words.

Spas in relatively remote places have attracted health-seekers throughout the world. In the USA, one of the objectives of the National Park Service is to provide spaces for a return to nature. Box 7.2 provides details on the success of one of these parks.

Box 7.2 Denali National Park, Alaska

Eugene Palka (1995b, 2000) has shown that a national park can be experienced as a therapeutic place because of the way visitors react to their natural surroundings. He chose Denali National Park in Alaska as a study site because of its spectacular scenery and its reputation for having preserved a pristine wilderness untouched by human intrusion. The idea that Denali is a healing place has been built up over the years. Dale Brown, a writer and explorer, wrote (1984: 103), 'I came to Denali tired, in need of change. What I found invigorated me, filled me with happiness as warm as sunlight.' Here is another reaction, from Sheri Forbes (1992: 31):

> [As] you follow the trail to the lake, the human-caused sights and
> sounds that surround you will be gradually replaced by the signs
> of nature, and you will enter a realm where the body can relax,
> the mind can expand, and the spirit can be restored.

Interestingly, although the park had over two and a half million visitors in 1991 and 1992, whose average age was 57, only eighty people received medical care. Furthermore, most of those treated were mountain-climbers and therefore not typical tourists. Because of the low demand for care, the park has no medical facilities; emergency treatment is provided by park rangers.

To find out how people experienced nature in the park, Palka (1995a) questioned over 160 visitors who took the long bus trip through the park's landscapes. People reported being enthralled by such natural features as the towering mountains, streams and rivers, and sunsets. Many mentioned the authenticity of the natural setting; others were most attracted by the varied and abundant wildlife. Some said that they had just wanted to get away from it all. Most visitors claimed to feel better after their trip through the park. Asked if the place had a therapeutic effect, people responded with phrases such as 'Denali is peaceful, harmonious, and relaxing', 'I was able to recharge my batteries', or 'My experience was therapeutic in many respects. It was refreshing, peaceful, tranquil, and restorative.'

It must be noted that the feelings for Denali's wonders expressed by visitors are influenced by factors besides what appears to be a human affinity for nature. People come to the park with high expectations that they will enjoy the place, having read about Denali and listened to the tales told by previous visitors. Park rangers and park literature further indoctrinate the willing visitor with their views about the benefits of untouched nature. As people ride along in the bus and get out to view the scenery, they become knowing insiders who reinforce each other's feelings about the sense of well-being the park produces. Thus nature and human imaginations conspire to construct a therapeutic atmosphere.

Selling places that heal

An important aspect of health care 'reform' in developed countries in the 1980s and 1990s has been the commodification of health care, driven by the market-based ideologies espoused by New Right politicians (Kearns and Barnett 1997). Health has increasingly become more a commodity than a quality and health care a product rather than a service. The recent emphasis on selling health is rooted in a consumerist ideology fostered by late twentieth-century capitalism and the condition of postmodernity. Consumerism 'emphasizes the good life for all, through the individual pursuit of objects to satisfy individual wants' (Eyles 1987: 94), creating the illusion of free choice and neglecting such societal concerns as equity.

Although we may be tempted to consider the commodification of health as a (post)modern phenomenon, it is probably safe to say that selling health care products for profit and selling people on buying them are as old as any medical system. Health is so vital to humans that they are often easily swayed by the (often dubious) claims of those who would gain by their vulnerability, whether they be purchasing snake oil or the newest in headache pills. The history of the selling of spas to the eighteenth- and nineteenth-century publics in countries such as England and the USA is an excellent case in point. During its glorious Georgian period Bath was sold as much for its architecture and glittering social life as for the curative properties of its mineral waters, although the city's reputation for healing was contested (Box 7.3).

Box 7.3 Constructing a healing reputation

Certain places have achieved a reputation for healing that has lasted for at least decades (e.g. the Mayo Clinic in Rochester, New York) and some-times more than a thousand years (e.g. Epidauros in Greece, where the god Asclepius healed in dreams). Positive feelings about the therapeutic value of a place are constructed from historical events, promotional efforts and the experiences of visitors; over time, the feelings become fixed as 'understood truths'. Here we briefly discuss some of the factors that contributed to the English city of Bath's reputation as the premier watering place of the nineteenth century (Gesler 1998).

The mineral springs at Bath lie at the heart of its therapeutic reputation. Celtic people worshipped their gods here and the Romans, who constructed magnificent baths, used the site for rest and recreation (Figure 7.1). During the Middle Ages, Bath attracted pilgrims from all over Europe to bathe in its holy waters, but it was during the Georgian period of the eighteenth

---

century that the town reached its height of popularity. Particularly important in contributing to Bath's fame were its symbolic landscapes (discussed
in more detail later in this chapter), including a healing myth about King
Bladud (healed of leprosy in the mineral springs), and the sacred buildings
(e.g. Queen Square, King's Circus) constructed by the master architect
John Wood (Cunliffe 1986; Neale 1973). Figure 7.2 is a photograph of
the Royal Crescent, built by his son.

Important as Bath was as a healing place, however, its reputation was contested,
especially during the Georgian era. Detractors said that the site itself was
unhealthy, and both physicians and novelists disputed the claims that the waters
had the power to cure illnesses. Hygiene was clearly a very serious problem in
the baths and in the streets. Many doctors were quacks who bilked the unwary,
and the poor (the Beggars of Bath became a social category) caused endless problems. Still, despite its detractors, Bath's healing reputation survived. What seems
to be the key to its success was that people perceived it to be a healing place
and so it became one in the public imagination.

Figure 7.1 The Roman Baths in Bath. These elaborate baths were built by the Roman
invaders in the 60s and 70s of the first century AD, became buried as the town
grew over them, and then were discovered and excavated in the 1880s as Bath
reinvented its glorious past. Photo by Lucy Gunning.

126

Figure 7.2 The Royal Crescent, Bath. Built by John Wood's son, John Wood the younger, between 1767 and 1775, this magnificent building, in Cunliffe's (1986: 134) words, 'fully deserves its reputation of being perhaps the most perfect essay in urban architecture in the western world'. Photo by Lucy Gunning.

Like most spas in their early days, the mineral spring resorts of colonial and newly independent America first attracted people who were mainly interested in restoring or maintaining their health. Gail Gillespie (1998) has traced the movement away from strictly health concerns to the selling of health and place in nineteenth-century spas located in the Upper South. In the decades preceding the American Civil War, the coming of the railway to the Valley of Virginia enabled people from the middle classes to visit spring resorts that previously had been frequented almost exclusively by the aristocracy of the old South. Entrepreneurs, bolstered by an economic boom in cotton, began in the 1830s to organize stock companies to build new resorts along the transport routes that served older, established springs. They employed artists to depict or represent these resorts in brochures and guides in ways that would appeal to prospective customers. Because the resort-owners belonged to the culture of their clientele, they understood what visual (as well as verbal) images would present a way of life that was most compelling. Their strategy was to select points of view, to manipulate perspectives and proportions, that would (re)present visual metaphors of the prevailing social order.

Although it was changing in the mid-nineteenth century South, the Southern social order was still symbolized by the rigid pyramidal hierarchy of power of the plantation. Plantation buildings, gardens, roads and paths were laid out to represent an ordered world in which power came from God and descended down through the plantation owner, his family and, lastly, to slaves (Vlach 1993). Advertisements showed spring landscapes that closely resembled those of the plantation. A specific device used by artists was the bird's-eye panoramic view, which showed the entire human-made landscape situated in natural surroundings and a very rigid and symmetrical arrangement of buildings. This served to symbolize the Southern social structure, which had both found its place within the natural world (and hence was 'natural' or God-given) and represented desirable stability, order and a proper hierarchy of power. Thus, '. . . the pictures of the springs produced through commercial interests in the middle of the nineteenth century were a visual coding of cultural values' (Gillespie 1998: 188).

'The history of Hot Springs, South Dakota,' Martha Geores writes, 'is the history of a town founded by entrepreneurs to sell a commodity called "health"' (Geores 1998: 36). She shows how this place was sold to generations of Americans from the late nineteenth century to the present through the use of a specific metaphor: 'Health = Hot Springs'. Clearly, entrepreneurs such as the Hot Springs Publicity Bureau and railway companies used the inherent ambivalence of metaphor (see Chapter 5) to promote meanings that changed as both individual and cultural definitions of health changed over time. In the 1870s and 1880s, profit-seekers exploited the mystique and exotic nature of Native Americans when they told potential visitors that the hot springs were a healing place sacred to the original occupants of the area (but suppressed the less-than-friendly relations between the US government and the Indians).

Another ploy used in Hot Springs was to extol the unusual local climate, which did not feature the extremely cold winters and oppressively hot summers typical of much of the Dakota lands. Promoters also changed the name of the place from Minnekahta to Hot Springs in order to benefit from 'naming as norming' (see Chapter 5) and association with the more famous Hot Springs, Arkansas, and Hot Sulfur Springs, Virginia. In imitation of English spas, Hot Springs was touted as an upper-class resort that barred disreputable people. The waters themselves, of course, were said to cure a broad spectrum of diseases. Notably, however, the list did not include contagious diseases; given that the USA was at the beginning of the epidemiological transition (when most people die from infectious and contagious diseases) at this time, this eliminated most ailments, especially those experienced by the lower classes. As time went on, attention became less focused on the mineral springs and more on the main resources of an emerging biomedicine, namely doctors and hospitals. In one astute move, railways sponsored medical conventions for doctors, and doctors in turn passed resolutions praising Hot Springs. The town eventually had far more than its share of hospitals, given the small population of the surrounding region, but they helped the town to survive. Promoters even managed to obtain the endorsement of both the US government and the American Automobile Association.

Symbolic healing landscapes

It is a human trait to use both concrete and abstract symbols to give meaning to one's life: a red cross expresses caring help for those in extreme need, the physician's white coat represents purity and health, and the idea that mentally ill people are 'deviant' is a socially constructed means of stigmatizing them. Symbols can be used to create identity (e.g. a nation's flag) or separation (e.g. slogans painted on walls in Belfast, Northern Ireland), and they can arouse intense positive or negative feelings (e.g. the Communist hammer and sickle). Figure 7.3 shows one of a myriad of uses of the Christian cross. Many symbols are ambiguous; that is, they can be interpreted in different ways at different times and places by different people. For example, a massive new hospital complex can inspire awe and trust in some, fear and distrust in others. Symbols pervade our everyday lives; in fact, some anthropologists feel that the study of culture is really the

Figure 7.3 Roadside crosses marking places where fatal car accidents occurred. For Christians, the cross is a potent symbol of death, but also points towards resurrection. The use of crosses in this situation may indicate both a warning and a memorial. Photo courtesy *Directions* magazine, New Zealand.

study of complex symbolic systems (Turner 1975). The landscapes we create are invested with symbols that express our beliefs and social relationships. As Meinig (1979: 6) states, 'We regard all landscapes as symbolic, as expressions of cultural values, social behavior, and individual actions worked upon particular localities over a span of time.' We 'read' or interpret landscapes to extract the symbolic meaning they have for various people.

In the realm of health, symbols 'work' because they mediate between people's biophysical and social worlds (Kleinman 1973). That is, meaning is brought to our experiences of disease and health through symbolic associations with our cultural and social life; a similar point was made in Chapter 5's discussion of metaphor. This essential point requires some illustration. In developing countries where malaria and diarrhoea are endemic and apparently random as well as highly malignant, people find an explanation for these diseases in a symbolic cultural complex that is equally random and malignant, namely witchcraft. In one society, faced with a difficult childbirth, a shaman tells the prospective mother a well-known story about a band of heroes who successfully complete a dangerous journey, stopping at places the shaman relates to places along the birth canal; the birth is successful, too (Dow 1986). The Ojibway of southern Ontario are at high risk for diabetes and alcoholism. It has been discovered that these problems can be alleviated if patients are told a myth that opposes the gluttonous Windigo, symbol of the symptoms of diabetes, with Nanabush the teacher, who represents truth. This myth is in tune with Ojibway beliefs that spiritual strength is necessary to good health (Hagey 1984). The remainder of this section further illustrates symbolic landscapes of health, detailing examples from three places: Havana in Cuba, Bath in England, and Iquique in Chile.

The government of Cuba is extremely proud of its health service and rightly so: as a developing nation politically and economically at odds with the USA, it has created a system of medical care that ranks among the best in the world (Feinsilver 1993). In the early 1990s, for example, Cuba's infant mortality rate (11.1 per 1,000 live births) ranked third among 191 nations. Cuba very consciously displays its success in health care to make the ideological point that socialism can produce a healthy population; this effort is clearly seen in the built landscapes of Havana. The sheer number of medical facilities is impressive: clinics, hospitals, research facilities and offices of family doctors can be found on almost every block, indicating that health care is always near at hand. One of the tallest buildings in the city is the Hermanos Amejeiras Hospital, named after brothers martyred in the revolutionary struggle Castro led against Batista. The Center for Genetic Engineering and Biotechnology is claimed to be one of the largest laboratories in the world. José Martí Pioneer city is a seaside youth vacation camp that can house around 10,000 children and 4,000 mothers; it achieved international fame when it expanded to take in hundreds of children injured by radiation fallout from the Chernobyl nuclear power plant in the former Soviet Union (Scarpaci 1999).

Other scenes in Havana contribute to the symbolic landscape of health under socialism. Although commercial advertising is now appearing in 'post-Soviet' Cuba, during most of the Castro period the only brightly painted objects in Havana represented political ideology; for example, messages from heroes such as Che Guevara or Raúl and Fidel Castro. Large signs also sell the importance of health care. As an example, on a wall in front of Calixto García Hospital, a sign reads, 'The health of one man is more valuable than all the land of the wealthiest man in the world.' Prominent, too, in Havana's built environment is health tourism, Cuba's development of sophisticated medical services for both foreigners and nationals. Several facilities (sometimes offering alternative, new or unproved practices that may not be sanctioned in other countries) cater to those seeking relatively inexpensive care and, at the same time, advertise Cuba's health care delivery successes (Scarpaci 1999).

Returning to the example of Bath used earlier in the chapter, we note that the place held symbolic meaning for the Celts, Romans, Normans and Georgians (Gesler 1998; Box 7.3). Two aspects of the symbolic landscape of Bath stand out: the myth of King Bladud and the architecture of John Wood. Bladud was said to have been exiled from his father's court because he had the heavily stigmatized disease of leprosy. According to legend he became a swineherd and infected his animals with his ailment. Then one day the pigs and he were cured by immersing themselves in the mud around the mineral springs. Bladud was welcomed back into the court and in time became king. This legend, with its mention of miraculous cures and royalty, helped to establish Bath's healing reputation. The tale itself can be interpreted in different ways, all of which enhance a healing theme. It may, with echoes of the New Testament story of a man being cured of his 'demons' when they were transferred to a herd of swine, refer to mental illness. Or it may be associated with the story Jesus told about the prodigal son and thus either refer to healing broken social relationships or give comfort to patrons of Bath's waters who alternately ate and drank too much and then sought remedies for gout and other ailments.

The eighteenth-century buildings that made Bath architecturally famous worldwide were largely the creation of John Wood, who had a vision of constructing a New Rome (Cunliffe 1986). Influenced by both pagan and Christian ideas and inspired by Andrea Palladio, a late-Renaissance architect who based his own work on Roman designs, Wood created several masterpieces, including Queen Square and the King's Circus. The plans for his buildings were based on two symbols, the circle and the square, which represented the Divine Architect's striving to produce harmony and perfection in the world. It is quite likely that Wood was strongly influenced by Renaissance views that held that the maintenance of order was necessary for physical and mental health, and thus the order and symmetry of his constructions were attempts to heal through architecture (Bamborough 1980; Mills 1982). In particular, his designs were based on the Vitruvian figure of the human body with outstretched limbs, thus linking architectural order to human balance and health.

Symbolic healing landscapes may take on many forms. In contrast to the magnificence of Wood's masterpieces, Lessie Jo Frazier and Joseph Scarpaci (1998) write of landscapes created in Chile to come to terms with the horrors of the repressive Pinochet dictatorship that lasted from 1973 to 1990. Pinochet was responsible for killing, by torture and assassination, thousands who opposed his military regime. Beginning in 1990, several mass grave sites throughout Chile have been unearthed, evidence of a topography of state terror (Miranda 1989). As this 'landscape of terror' was revealed, the human rights movement in Chile focused attention on the grave sites, and communities gathered to hold funerals and mourn for the dead. Visual documentation of the opened sites appeared in posters, murals, magazines and documentaries, and allowed Chileans to 'read' the history of violence inscribed on the bodies of victims. Human rights groups tried to make sure that the truth would come out, that the past would not be forgotten. Thus an attempt was begun to turn a landscape that symbolized a pathology of state violence and repression into a landscape of reconciliation and healing.

Associations of former political prisoners and the families of disappeared or executed persons in Chile demanded reparations (which included health care for victims of state terror) from the new democratic government that replaced Pinochet. Iquique, the primary port city in the northernmost frontier province of Tarapacá, was chosen as the site of a pilot reparation programme called Mental Health and Human Rights. This health care programme provides treatment to families of executed and returned exiles and former prisoners. The choice of Tarapacá as a *place* for this effort was especially significant as it played important roles in the forming of the Chilean state. It was the scene of struggles over nitrate wealth, the birthplace of the Chilean labour movement and most of the contemporary political parties, and also is where Pisagua, a detention camp for political prisoners and homosexuals, is located. Iquique represents the ways in which a community 'struggles to reclaim its spaces and well-being through protest, testimony, and the search for mental health and integrity' (Frazier and Scarpaci 1998: 56). Victims of state violence, as well as its perpetrators, can potentially be healed by reconciling the pathologies and therapies that occurred in the same place.

Therapeutic landscapes

What makes a place a healing place? In attempting to answer this question, this chapter has emphasized the importance of different landscapes that affect people's experience of healing. We examined natural, consumer-oriented and symbolic landscapes, but others could be considered as well. It is necessary to be eclectic when examining therapeutic landscapes simply because a wide variety of influences on the healing power of place exist. Anthropologists, for example, focus on the landscape of belief, although they do not use this expression, and sociologists study landscapes of social relations. In other words, examining therapeutic landscapes is an interdisciplinary venture. It is geographic in that it deals with specific places, but it brings together layered landscapes of meaning from several sources.

Gesler (1992) has attempted to categorize landscape possibilities, realizing that it is an imperfect effort at best. Two points about such a categorization or therapeutic landscape model should be made. The first is that landscape types are interconnected and overlap. For example, a model might distinguish between natural and built environments, but humans build nature (e.g. a flower garden) into their healing places. Second, every healing place tells its own story; the relative importance of various therapeutic landscape types will vary from situation to situation: a natural environment might not be as important in a modern hospital as it is in a country retreat for those dying of cancer (although perhaps it should be). The remainder of this section describes three very different situations that illustrate combinations of types of therapeutic landscapes.

This first study is of health camps for children in New Zealand, first established in 1919; currently seven camps serve approximately 4,000 each year. Kearns and Collins (2000) show how perceptions about what is therapeutic about the camps are historically contingent, depending on changing ideas about children's health (see also Thurber and Malinowski 1999). In the period following the First World War there was a strong interest in countries such as Britain and New Zealand in national fitness and improving the white race through eugenics. Camp life was strictly regimented, mirroring military life. At the same time, camps were set up to combat such ailments as tuberculosis and malnutrition. Emphasis was placed on the healing powers of fresh air, exposure to sunlight, and large meals. From the 1960s, the original focus on physical well-being began to shift towards more of a concern for the psychological aspects of children's health, and attention was paid to children with behavioural problems and those from dysfunctional homes.

The New Zealand health camps have always been supported by a mix of private and public funds, but over time the responsibility for funding has shifted more towards the state. Under the conservative National Party government from 1990 to 1999, this meant an emphasis on efficiency, contestability and accountability. Camp managers were required to make contracts with funding agencies and meet certain performance targets, as well as compete for money with other health and welfare agencies. At the same time, there has been a strong movement to deinstitutionalize child health care and focus on treating children at home. Nonetheless, the camps have remained open. Managers and supporters continue to extol their therapeutic qualities, and indeed there remains a feeling that they are a significant part of the national heritage. Although they are under severe pressure, there is a growing demand for places in the camps.

Rebecca Dobbs (1997a, b) examined the therapeutic landscape of the Navajo, a Native American people who live in the south-western USA. This study differs from other studies of therapeutic landscapes in three ways. First, it deals with a non-Western society whose culture (e.g. in attitudes towards land ownership) may be very different from Western cultures and thus must not be approached with ideas derived from studying Western societies alone. Second, the therapeutic landscape is the entire Navajo homeland, not just a specific place or facility. And, third, the study deals not with visitors *to* a healing place but with the people who actually reside *within* the landscape. Thus we are dealing with a group of people who identify intimately with their various landscapes and consider all of them to have healing power. Let us look at five types of therapeutic landscapes that exist within the Navajo healing world.

The *natural landscape* of the Navajo homeland is noted for its beauty, and thus its potential restorative powers, by both Navajo and non-Navajo alike. Of particular importance are four sacred mountains that can be identified on the actual landscape and that figure prominently in Navajo healing mythology. Within the *built landscape* of the Navajo, the hogan or dwelling assumes the utmost healing importance. Its design follows a sacred, cosmic pattern and is the centre of ceremonial healing activities (e.g. the construction of a sacred sand painting). The Navajo have also created a *symbolic landscape* of mythic stories that trace the journeys of supernatural beings. These stories represent a sacred geography as they provide a vast amount of information about the places where supernatural events have occurred. Concepts developed about disease and its treatment provide a *landscape of belief*. As an example, the Navajo disease taxonomy can be seen as a nested hierarchy, beginning with sickness in humans, moving out to sickness in humans and animals, to all living things except plants, and finally to all elements of the Navajo universe. Finally, the Navajo *landscape of social relations* is illustrated by the social ties that are strengthened as many people outside the immediate family attend healing ceremonies performed for a family member.

Many of the landscapes discussed in this chapter may seem far from the reader's everyday concerns about health. What do the natural landscapes of ancient Greece, selling the virtues of nineteenth-century spas, combating the terrors of Chilean dictatorships, or Navajo stories of supernatural healers have to do with the places we encounter when seeking care? It is our contention that all these kinds of therapeutic landscapes can be discerned in modern health care systems. To demonstrate this, we turn now to an examination of some of the influences on a modern hospital of the same types of therapeutic environments as discussed for the Navajo (Gesler *et al.* 1998).

Although many people have a difficult time thinking about how *nature* plays a role in the modern hospital, many staff and patients find pleasure, and thus mental healing at least, in such things as fountains or gardens. A handful of experiments have shown that exposure to nature contributes to physical healing as well (Box 7.4). The *built environment* of hospitals – overall construction, individual

---

Box 7.4 The restorative effects of viewing nature

Roger Ulrich (1983) argues that if people confined over long periods in such places as hospitals, prisons or stressful work environments are exposed to views of nature, this may have a restorative effect. This argument stems from *prepared learning theory*, which holds that evolution has predisposed humans to easily learn and maintain associations or responses that aid in their survival (Seligman 1970). Thus humans often react negatively to aspects of nature that may have been risks in the past (e.g. snakes, forests) (*biophobia*) and positively to such things as savanna landscapes (open areas where early humans evolved and could view potential enemies easily) and water (the staff of life) (*biophilia*). Several studies have shown that different groups of people from Europe, North America and Asia prefer to look at rural as opposed to urban views. Scenes with vegetation, water and mountains are preferred over those with buildings, cars or advertising signs. Views with irregular or curvilinear contours or edges are chosen over those with straight or regular lines.

A few studies have provided direct evidence of the power of viewing natural settings to restore people in stressful situations. Examples include exposing patients in health care settings to nature for short periods (e.g. ten minutes), showing ceiling pictures to patients lying on a hospital gurney, and studying the effects of different types of wall art on psychiatric patients. In a classic experiment, Ulrich (1984) separated patients recovering from gall bladder surgery into two groups matched for variables that could influence recovery such as age, sex, weight, tobacco use and previous hospitalization. Pairs of patients were assigned randomly to rooms in the hospital. One member of each pair had a view from a window that looked out onto a grove of deciduous trees, while the other patient looked out at a brown brick wall. Ulrich found that the former patients had shorter postoperative stays in the hospital, fewer negative comments in nurses' notes, lower scores for post-surgical complications such as nausea and headaches that required medication, and less need for painkillers.

Box 7.5 Ambiguous perceptions of an asylum

We have spoken elsewhere in this text about the transition in the treatment of those with mental illness in large institutions (asylums) to deinstitutionalization or 'care in the community'. But what happens to the old asylums? Do they lose their meaning? Claire Cornish (1997) set out to discover what was happening to St Lawrence's Hospital in Bodmin, Cornwall, founded in 1820. Currently this institution is being 'run down'. Moves to rationalize the use of space within St Lawrence's have created a state of flux, a transient micro-geography. A population that once was estimated to be over 1,500 is now reduced to approximately 230 and once overcrowded buildings are now silent. Facilities and services are being shifted from the oldest buildings in the 'top end' to new buildings at the 'bottom end'.

Cornish has identified two contrasting meanings that St Lawrence's has for people today. Traditionally, contacts between the asylum and the community were kept to a minimum and as a result the place was stigmatized out of ignorance about what went on behind its perimeter fence and massive iron doors. For many, two centuries of harbouring the dreaded and misunderstood 'other', the deviant from normal society, the misfit, has perpetuated a strong sense of stigma. Some of the unused land within the asylum could be used for other purposes, but people are reluctant to deal with tainted space. In contrast, many people, including staff and Bodmin residents, think nostalgically of the place as a community, a home for patients, and a place where people were united in a common purpose. These people opposed the running down of the asylum and said that putting patients into the community was a negative step. Thus the meanings attached to St Lawrence's Hospital are both varied and ambiguous.

room design, lighting, ventilation, cleanliness, noise levels, colour schemes, and so on – have been considered in many studies (see, for example, Hutton and Richardson 1995; Williams 1988). Many of the concrete objects or artefacts that patients and staff encounter may have *symbolic meaning* for them: a physician's white coat may stand for purity or honesty (Blumhagen 1979); and high-tech equipment can either frighten or impress (Kenny and Canter 1979). Sometimes, however, the meanings of a place may be ambiguous (Box 7.5).

Some hospitals consciously construct images that are supposed to attract and soothe patients; a good example is Auckland's Starship children's hospital, which features objects relating to outer space and adventure (Kearns and Barnett 1999). Various actors bring different *beliefs* about disease and its treatment to medical encounters with hospitals; the degree of agreement of explanatory models (see Chapter 5) among patients and staff can have an important effect on healing (Mishler 1984; Todd and Fisher 1993). And, clearly, *social relationships* among and within staff and patients in hospitals can create landscapes that either help or harm. Some of the issues involved here are dominance hierarchies among staff, patient autonomy (Kenny and Canter 1979), and the quality of interactions among staff (Leatt and Schneck 1982).

Landscape/place/healing

In this chapter, landscape was used as a heuristic device to examine the environments of places in which health care takes place. We began by taking three perspectives on landscape – human–environment interaction, social construction and mental construct – and applying them to health-in-place issues under the rubrics of nature as healer, selling places that heal, and symbolic healing landscapes. These three approaches are derived from the cultural ecology theme of more traditional cultural geography and structuralist and humanist theories from the newer cultural geography. In the last section of the chapter we introduced a five-part taxonomy that partially overlaps but also extends the original three-part scheme: natural, built, symbolic, belief and social relations landscapes. No doubt other landscapes could be employed to describe healing places, but these three or five have been found to cover many situations, from an Asclepian sanctuary in ancient Greece to a modern hospital. One cannot measure in quantitative terms the relative influence of the different landscapes on any one healing situation – some will be more applicable in one place and time than others – but it is probably worth looking at them all. In fact, they work together in a synergistic fashion and thus their effects are difficult to separate. Improvements in one or more of them can potentially lead to better health.

We think of landscapes of healing as an ideal, a potential, something to strive for. Of course, what is ideal for one person is not for another, so we must allow for difference. Furthermore, the healing qualities of a place may always be contested. Ideals are never achieved, but we argue that knowing what their parameters might be adds to our knowledge about health care.

Further reading

Kearns, R. A. and Gesler, W. M. (eds) (1998) *Putting Health into Place: Landscape, Identity, and Well-Being*, Syracuse: Syracuse University Press. This selection of readings by health geographers emphasizes the issues or concepts of therapeutic landscapes, identity and difference, and health policy.

Canter, D. and Canter, S. (eds) (1979) *Designing for Therapeutic Environments*, New York: John Wiley. This book deals with the ways environmental psychologists think about how built environments can be produced that are conducive to health and well-being.

Meinig, D. W. (ed.) (1979) *The Interpretation of Ordinary Landscapes*, New York: Oxford University Press. This classic contains essays on humanistic approaches to viewing the landscapes that we experience in our daily lives.

Williams, A. (ed.) (1999) *Therapeutic Landscapes: The Dynamic between Place and Wellness*, Lanham, MD: University Press of America. Various themes arising from the therapeutic landscape concept, including healing places, marginalized people and health care systems, are written about by health geographers in this collection.

8

CONSUMPTION, PLACE
AND HEALTH

Introduction

Until the mid-1990s, geographical analysis of health care consumption was often undertaken by examining patterns of service utilization. Furthermore, the sites of service provision were viewed as locations rather than as contributors to, and constituents of, health care landscapes (e.g. Joseph and Phillips 1984). In this chapter we explore ways in which ideas linking landscape, language and identity surveyed in the preceding chapters can assist in interpreting the evidence of consumption landscapes of health care. This idea was first introduced in Chapter 7 under the heading 'Selling places that heal' (p. 125). According to Eyles (1987: 94), consumerism 'emphasises the good life for all' and the myth of consumerism holds that we are all equally free to choose, even though people cannot in actuality all enjoy the same type of product.

Consumerist ideology has only recently been incorporated into health care services. We take 'consumption landscapes' to be the material expressions of appeals to 'consumers': those who are targeted to buy, acquire or use goods and services. Examples such as accident and medical clinics, a new private hospital, and the use of malls and fast food outlets by psychiatric patients serve to illustrate how underlying ideologies of competitive provision are reflected in the symbols used both in the built environment and in advertising.

While our examples are drawn from health systems we are familiar with, our argument is that no matter what the health care system, consumerism plays a key role. Indeed, beyond the tangible built environment, the quintessential consumer's world of television is being increasingly colonized by medical landscapes through the popularity of shows such as *ER* and *Chicago Hope*. What is consumed on television is not health care or even place, but rather the drama and perceived heroism of medicine itself. If much advertising relates to the 'magic of the mall' (Goss 1993), these consumer's worlds celebrate the 'magic of the ward' and feed into a view that privileges biomedicine over basic public health interventions. Thus, in our view, an interpretation of health care sector reforms in Western nations can be undertaken not only through the more usual assessment of policy and outcomes (e.g. Baggott 1994), but also through a reading of the texts they produce. These texts, we contend, comprise the various means by which messages are sold – and include both policy documents that emanate from governments and their agencies (Moon and Brown 1998) and, of interest in this chapter, the specific ways in which health care enterprises project themselves in the landscape.

We organize the remainder of the chapter as follows. First we review the character of consumption, before considering the changing appearance of consumption landscapes. Specifically, we describe private accident and medical clinics that have been developed contemporaneously with state health care reforms in the 1990s. A third section examines the links between symbolism, landscape and health care. The fourth section explicitly considers the presence of consumerist ideology in the landscape, and a final substantive section examines the ways people who are less able to purchase goods and services nevertheless utilize consumption landscapes in perhaps unexpected ways.

The character of consumerism

To consume implies sets of practices and perceptions that are predicated by, and embedded within, economic and power relations. First, to consume implies the presence of something consumable. Buyers imply sellers. However, the simplicity of these relations, typified by negotiations at a village marketplace, is made more complex and ambiguous within contemporary society. A common feature of health reforms in Western nations has been an introduction, to variable degrees, of the market-based ideology advanced by so-called New Right politics. This shift has resulted in an infusion of competitive practices and greater levels of advertising into health care. An outcome of this process has been the construction of health and health care as commodity and product as well as quality and service. Hence, we could argue, the health care system itself has become part of what Robert Sack (1992) describes as 'the consumer's world'.

Second, to be a consumer is to participate, and be part of 'community involvement' in health service planning and provision. As Curtis and Taket (1996) imply, this culture of participation is fraught with ambiguity, for there is a fine line involved in striking the 'proper' balance between power being vested in government and people respectively. What may seem like the power of self-determination (e.g. as 'mental health consumers') can sometimes be unmasked as attempts by the state to reduce its role as a welfare provider. This possibility of consumer participation being either a New Right attempt at self-provisioning or a left-of-centre concern with rights and self-help by service-users leads Croft and Beresford (1992) to call the former 'consumerist' and the latter 'democratic' approaches to user involvement. There are differing bases for these approaches. Curtis and Taket (1996) describe consumerists as being concerned with costs, service quality and the role of the state in perpetuating dependency whereas democratic concerns with involvement originate more in equal-opportunities struggles.

Perhaps another way of thinking about these dual uses of the 'consumer' idea is the distinction between individual and citizen consumers. Individual consumers are appealed to in the free market of signs and advertising with the implicit exhortation to look after their own needs through securing the best deal possible. Alternatively, citizen-based consumerism buys into participation, seeking the reward of gaining a stake in health services.

In this chapter we mainly focus on evidence in the landscape of the individual consumer's world of health care whose roots lie in the political ideology associated with twentieth-century capitalism. We can explain the necessity of bold promotion of cure or care (rather than simply healing reputation, as discussed in Chapter 7) to competitive practices. Thus advertising, by means of road sign, distinctive architecture or printed slogan, was adopted once there was an abundance of practitioners, competition for patients, and the presence of clinics no longer associated with a particular well-known doctor.

The changing face of consumption landscapes

There has been a rapid and profound shift in the style of consumption over recent decades in Western countries. To a considerable extent, corner shops as well as downtown department stores have been displaced by centralized and internalized malls containing a host of stores and other attractions. In turn, the idea of the mall has 'colonized' other places in the city, including sites of health care provision. It is now common to find cafés and retail outlets in hospital foyers, as well as mini-medical malls amalgamating a range of health-related services (Kearns and Barnett 1992). This shift has occurred at the same time as the act of both retailing and seeking medical advice from home has been increased through technology such as the Internet (Parr 2001).

Generally speaking, seeking health care, like shopping, is a thoroughly spatial activity: we go to the doctor because medical transactions conventionally take place in particular places. Following Kearns et al. (2001), we can view consumption as an ideology, attitude and behaviour that can be connected to three key processes associated with the postmodern era (Lash and Friedman, 1992):

1   The *increasing commodification of social life*, such that signs and images increasingly 'sell' consumption and intrude into the processes by which our identities are shaped. In other words, it matters to teenagers whether their running shoes are Nike (Skelton and Valentine 1998). Similarly, it matters to some that they hold private health insurance, not only because it allows ready access to elite facilities but also because the very eliteness of such clinics and hospitals affirms and confirms their identity. In New Zealand, for instance, 'gold' and 'silver' insurance plans available from a major company epitomize this sense of eliteness and distinction, and echo the classes of frequent flyer programmes.

2   *Blurred social divisions* in society. A social implication of consumption is that the meanings of commodities and services in capitalist society become unstable, so that it is less easy to identify a person according to rank or class. Thus, advertising and fashion become tools in an attempt to stabilize meanings and create 'structures of taste'. The proliferation and diversification of gymnasia in urban places provides an example of places differentiated by taste. High membership fees at some ensure that membership is economically homogeneous and make them places of exclusivity.

3   *New forms of everyday life* are generated via consumption, with an increased use and reliance upon created spaces of consumption such as malls, museums and theme parks (Shields 1992). For example, even in summer, families flock to new suburban aquatic centres in Auckland as an alternative to local beaches. With the expansion of the ozone hole, shifting perceptions of sun safety, and pollution of urban beaches, such centres could be construed as safer places. Amid contrived landscapes of plastic palm trees, mock pirate ships suspended from the ceiling, and artificial waves, such places allow the display of clothes, and even of bodies. Yet the healthiness of these crowded, chlorinated spaces is debatable. We can draw a parallel with fantasy-spaces within hospitals such as Auckland's Starship children's hospital (see Kearns and Barnett 1999) and note how common the blurring of distinctions between representation and reality has become.

The interaction of these three processes can be identified in the links between place, identity and consumption in everyday life (cf. Glennie and Thrift 1992). The promotion of a culture of (health care) consumption has thus become evident both in printed media and in the visible landscape. With respect to the latter, there has been a greater prominence afforded to signs and to the buildings themselves that house health care activities. Paradoxically, as the balance of responsibility for health care has shifted from the public to the private sector through a restructuring of the welfare state, health care provision has moved from private, interiorized settings to more public and prominent places.

'Walk-in' clinics sited within shopping malls provide a good example. This trend amounts to a symbolic shift in the locus of health care from the world of the service user to the world of the consumer.

While a number of geographers have analysed the relations between place and consumption ideologies, health care is conspicuously absent from this work (e.g. Jackson and Thrift 1995; Gregson 1995). However, as Box 8.1 suggests, new forms of health care provision have taken highly visible forms in the built environment of cities. These developments are supplementing other international trends such as gentrification and shopping malls in contributing to a broader transformation of the cities into more seductive landscapes of consumption. These elements of the built environment contribute to the commodification of health through the activities of developers and medical entrepreneurs.

What do consumers of postmodern health care think about their treatment? Kearns and Barnett surveyed 205 patients in two accident and medical centres, one in inner-city and one in suburban Auckland, to find out. More than one-fifth (22.3 per cent) of the respondents said that aspects of the ambience of the clinics were what they liked; for low-income patients the percentage climbed to 65.9 per cent. For these respondents, form rivalled function in importance; ambience clearly can be sold. This result should not surprise us, however, as Singh (1990), writing in the *Journal of Health Care Marketing* (note the title!), states that there is a growing trend towards the 'marketization' of the health care industry. In a similar vein, Hutton and Richardson (1995) state that place is part of the 'total product' and 'atmospherics' affect the 'purchasing propensity' of buyers (potential patients).

Symbolism, landscape and health care

Doctors as well as clinics advertise their services. One promotional flyer discussed by Kearns and Barnett (1997) advertises a group of family doctors and offers one-stop services. It also promotes an image of contented consumers and doctors who are established within and serving faithfully their local community. Another example is a road sign for a clinic that incorporates both traditional general practitioners and non-traditional therapies under the collective title of the Holistic Medical Centre (see Figure 8.1).

Signs such as that depicted in Figure 8.1 contribute to the symbolism of health care in the city. Although in popular culture cities are frequently depicted in a negative light through the use of phrases such as 'the urban jungle', the city can alternatively be conceived of as a 'moral universe' in which strangers offer each other opportunities to seek care and comfort (Tuan 1988). Rarely, however, is such care offered without some sort of cost. Foucault (1976) pointed out that the very 'gaze' of professionals transforms persons into patients or clients and

Box 8.1 The malling of medicine

New Zealand is one of those Western countries that has experienced the recent commodification of health care as the market principles of late capitalism have made strong inroads into its economy. Just how this commodification is occurring is the subject of a study by Kearns and Barnett (1997). Using a semantic approach, they 'read' the health care landscape and interpreted its meaning for the health of New Zealanders. In particular, they have investigated the selling of accident and medical clinics in Auckland. Their focus is on three aspects of the current medical system: the built environment of private care, advertisements used to attract customers, and the perceptions of patients.

In 1986 New Zealand passed the Commerce Act, which encouraged consumers to shop around for health care and allowed doctors to advertise their services. One result of the new legislation was the establishment of accident and medical clinics. These clinics advertised through the radio and local newspapers and by mailing business cards and refrigerator magnets to private homes. They are also highly visible in the urban landscape, strategically located at consumer shopping hubs, along major shopping streets and in shopping plazas. Here they take advantage of close proximity to fast-turnover retail stores and restaurants and take on some of the characteristics of their new neighbours: fast service, discount prices and multiple types of care within one clinic building. Thus medical clinics take part in the postmodern turn in both architecture and popular culture (see Chapter 2).

Where once the medical profession was reluctant to advertise its work, a new competitiveness has fostered a rash of blatant efforts to attract customers in places such as Auckland. For example, a newspaper ad for a clinic features a picture of Uncle Sam (involving associations with US culture) pointing a threatening finger and saying, 'Don't let the flu get you.' Offering flu vaccinations at a reduced price, the notice also states that 'The flu will kill a number of over 60 year olds this winter. Many others will be ill for months', and 'Younger adults will be off work, unable to play sport, and too sick to party – maybe for weeks!' (see Figure 8.2). The latter threat certainly is aimed at the consumerist culture.

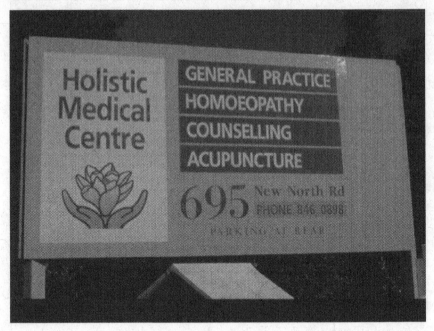

Figure 8.1 Branding of a primary health care clinic as inclusive of biomedicine and alternative therapies in Auckland. Note the symbols: flower and open hands. Photo by Robin Kearns.

this transformation involves a cost in terms of transformed identity. Other ways in which costs are incurred include monetary price (in the case of private care) or in potential losses of privacy (in the case of the crowded waiting rooms offered by 'welfare' health care). Another route into investigating consumption, health and place is through asking how ideas of private and public are expressed in the symbolic landscape of the city.

Hospitals are a good starting place for such an investigation. These sites of institutionalized medical care have tended to be highly functional elements in the urban landscape. They have generally advertised themselves only inadvertently through prominent location (e.g. elevated sites), or through their size (relative to surrounding buildings). The built form of many hospitals constructed in the latter half of the last century is austere and angular and their interiors potentially generate feelings of 'placelessness' (Relph 1976). In a similar functionalist approach, primary care clinics have often been located within office blocks or converted residential dwellings. However, influenced by the postmodern turn in architecture and popular culture, new for-profit clinics and hospitals proclaim their presence through 'place advertisement' techniques that involve 'architectural

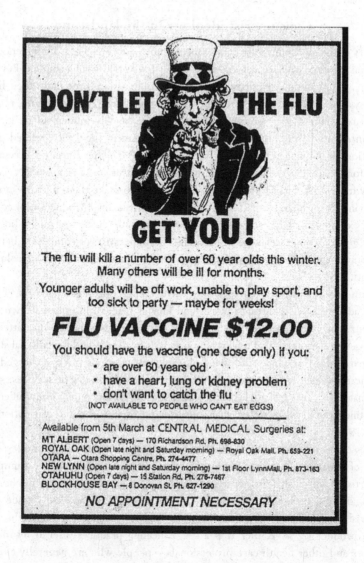

Figure 8.2 Newspaper advertisement for Central Medical Clinic, New Zealand, 1991. Source: *New Zealand Herald*.

imagineering' (Mills 1993: 152). As Box 8.1 indicated, this commodification is becoming accentuated through bold architecture and signs, mimicking the practices of other commercial enterprises.

Consumerist ideology in the landscape

As part of the sweeping shift from concerns for equity to those of efficiency that has characterized welfare reform in the Western world (Robinson and LeGrand 1993), a range of countries have encouraged competition throughout their health care system. For instance, both the British and New Zealand health reforms (Day and Klein 1991; Ashton 1992), have been heavily influenced by ideas of consumerism. Changes in the New Zealand system in the 1990s echoed policy directions contained in the British White Paper *Working for Patients* (Secretary of State for Health 1989). There were dual aims: to ensure greater financial accountability from providers of primary care, as well as to make them more responsive by having to compete for patients. In both countries, this introduction of competition extended into the secondary care system, with clusters of public hospitals being given a commercial orientation and the imperative of competing between themselves and with other (private) providers of secondary care (Malcolm and Barnett 1994).

With the imperative of competition, medicine, as a profession and set of practices, advertises itself through words and images that are inscribed in two types of texts: the landscape and the printed media. With respect to the latter, we proposed in Chapter 3 that advertising provides fertile ground for cultural studies and involves the practice of drawing meaning out of goods that are charged with significance. Such 'goods' are generally material commodities or services, so the fact that health (care) is advertised confirms that it has become commodified into something purchasable rather than remaining a quality and set of relationships (Epp 1986).

This commodification of health epitomizes the condition of postmodernity: a state of creative capitalism encouraging a culture of advanced consumption. Advertisers and health care entrepreneurs place a media-immersed public in an ambiguous position: they are socialized to be sceptical of the claims made about material products such as lawnmowers or washing machines, but health care is not a product *per se*. Rather it is a set of healing practices offered by medical doctors and other health care professionals – people who are generally afforded great respect in Western society. Therefore, although promoting health care *products* such as over-the-counter pharmaceuticals may be regarded with scepticism, the messages embodied in advertisements for medical *care* are likely to be considered more seriously by consumers. Symbols such as the white cross or even the word 'doctor' are messages replete with meaning. As we implied earlier, faced with acute pain or a family injury, a patient may be neither ready nor equipped to navigate the subtle variability of standards brought about by competitive practices and 'shop around' as a consumer (Barnett and Kearns 1996).

While most conventional general practitioners have been reluctant to advertise their fees, those entering the field of competitive primary health care sometimes show few scruples in this respect. A recent advertisement for a 'men's clinic' in Auckland's major newspaper, for instance, epitomizes the goal of advertising itself: 'to make us feel we are lacking' (Williamson 1978: 8). In the advertisement, a famous New Zealand athlete of the 1930s, Stan May, stands poised and ready to throw a javelin. However, instead of having the usual spear-like form, the javelin is wilted and drooping. There is an apparent incongruity in mixing an image of an Olympic athlete with a message promising an impotence remedy. Yet this image epitomizes the observation that in advertising, 'images from quite different contexts can coexist without contradiction because the message is not being

---

Box 8.2 Consuming elite health care

The culture of conspicuous health care consumption is evident in the Ascot Hospital in Auckland, which opened in 1999. In a city already well endowed with private hospital beds, the Ascot needed to rapidly establish a niche for itself. The chosen method was construction of an identity based around an elite site of unparalleled innovation and comfort (Kearns and Newman 1999). The hospital features sixty-eight individual in-patient rooms situated on its top floor. Each room has individual climate control, a personal telephone, cable television, a radio, coffee-making facilities, a refrigerator and an electronically controlled bed. If accommodated in a room (referred to as 'hotel space') at the rear of the hospital, a patient can enjoy spectacular views of the adjacent Ellerslie racecourse, a facility associated with class and sophistication. This ultra-modern (and expensive) place of post-surgical convalescence reflects attempts to offer a contrived therapeutic landscape (Gesler 1992) (literally) above and beyond the operating theatre. Ascot hospital management promote the idea that its ambience and the standards of service combine to be conducive to healing and recovery (Ascot Integrated Hospital 1999), a claim also made within the health care management literature (Hutton and Richardson 1995). The vast majority of 'consumers' of the Ascot are covered by health insurance, something only about 40 per cent of New Zealanders possess. The personal responsibility implied in this coverage is generalized in Ascot literature to suggest that patients 'take ownership of their treatment, supported by a clear exchange of information and liaison' (Ascot Integrated Hospital 1999). The patient, this view insists, is a consumer empowered by knowledge and explanation (Kearns and Barnett 1997).

These ideas were highlighted in publicity materials distributed at an open day held at the Ascot in early 1999. This event was advertised as '. . . an opportunity to see this magnificent new hospital and its amazing new technology' Potential consumers were enticed by 'Tiny laparoscopic cameras, powerful magnetic resonance imaging machines. . . the miracles of reconstructive surgery' (Ascot Integrated Hospital, 1999). With the air of a science fair, Ascot appealed through its technical wizardry. As if anticipating cynicism by some, compulsory donations were sought for Red Cross's first aid education in schools and the 'Books in Homes' programme. We can read involvement in these charities as helping to reinforce the Ascot philosophy of self-help. Soliciting money for these worthy causes in aid of children offered legitimacy to what was essentially the promoting of a business.

communicated as a "rational argument"; they are meant rather to evoke the realm of "meaning" ' (Leiss *et al*. 1986: 60). By way of example, the newly opened Ascot Hopsital in Auckland speaks, in its naming, to an English sophistication associated with fashion and horse racing and a sense of 'class' connected to the affluent part of Auckland in which it is located (see Box 8.2).

Finding space for non-consumers

In Western countries an increasingly conspicuous culture of consumption has developed, especially in urban areas. This process, combined with desire for a more cosmopolitan society, fuelled the development of New York-style cafés, gentrification of inner-city neighbourhoods, and the trend towards interiorized shopping malls. The culture of consumption has displaced conventional occupants of inner-city areas including long-term psychiatric patients in cities such as Vancouver and Auckland, through demolition or conversion of boarding house accommodation (Kearns and Joseph 2000). A good example is the Tuatara, one of the more expensive and exclusive cafés in Auckland. This café was developed in the refurbished premises of a drop-in centre for psychiatric patients in the 'trendy' inner-city suburb of Ponsonby and is now situated next to one of the accident and medical centres described in Box 8.1. As Kearns and Barnett (1997) discuss, the fact that respectable dining and entertainment establishments are juxtaposed with the clinics emphasizes the degree to which health care has been normalized and commodified in the consumer culture of contemporary Auckland. Box 8.3 describes how some users of such drop-in facilities cope in the face of a 'takeover' by the consumption landscape.

Box 8.3 Use of consumption spaces of psychiatric patients

Consumer-oriented landscapes can sometimes serve unexpected purposes for people with health care needs. This observation is especially the case when rationing of services, or the incomplete translation of a health care philosophy into a network of facilities, has occurred. A good example is provided by Caroline Knowles (2000), who describes the role that fast food outlets play in 'the fragmented task of daily survival' (p. 221) for mental health care clients (note the consumerist term!) in Montreal. She found that chains such as Burger King and Dunkin' Donuts have appeal to 'clients' as they are more ubiquitous in the city than day centres and other facilities, and they are open for longer hours. Furthermore, she claims that their popularity in the 'post-asylum era' makes them emblematic of a broader 'downloading' of people formerly managed by the health care system. Malls, the quintessential consumer's world (Goss 1993), are also identified by Knowles as places where mental health consumers spend time despite the fact that they often cannot, and do not, consume goods and services. Given that these apparently public spaces are in fact privately owned and managed, Knowles shows that their use by people with mental health problems is often contingent on the goodwill of security staff. Because one must consume to have a legitimate presence in the mall, she describes game-playing such as sitting in the food court with what looks like a coffee but, in fact, is a cup of water. As non- (monetary) consumers consuming the warmth and safety of the mall, mentally ill people have to attempt invisibility as the required price of participation.

Conclusion

Until recently, health care services have not been particularly visible within the landscape. While hospitals were prominent because of their sheer size, other services simply blended in, or were hidden from view. Recent health care reforms and the market-orientation of service provision in capitalist countries have resulted in competition between providers for both staff and patients becoming more intense. One outcome of this process is that existing providers not only have become more conspicuous in the built environment, but also have used a variety of other forms of advertising in an attempt to convey particular images of clinics and hospitals as appealing places of consumption. This trend has been stimulated in part by the entry of new competitors into the 'medical marketplace', but also by a growing belief on the part of new and existing providers that health care consumers can be potentially influenced by marketing strategies.

While there is a prevalent culture of consumption in Western countries such as Britain, the USA and New Zealand, we have argued that it is inappropriate to assume that patients will necessarily act in a consumerist manner. Thus, we have argued for an expansion of the idea of consumption within the geographies of health and health care. Models of consumption in geography have often been treated in sterile, asocial and historical ways (Warf 1994). The geography of health care consumption is no exception. Geographers have considered who uses particular services and why factors such as proximity have been more important to some groups of patients than others. But, as Sack (1992) has argued, we need to go further than this to understand the relationships between consumption and place, and why particular images and patterns of behaviour are characteristic of different groups of patients in these places. Consumption, therefore, may be perceived to involve more than just the purchase of medical care by particular groups of users. Rather, it may involve the ways in which messages about health are received and even rejected. By way of example, the 1997 opening of a McDonald's restaurant in Auckland's Starship children's hospital met with mixed reaction, including postcards anonymously placed on the windscreens of parked cars (Figure 8.3).

While we have focused on consumption landscapes in capitalist countries, links between consumption, health and place can be drawn in non-capitalist contexts

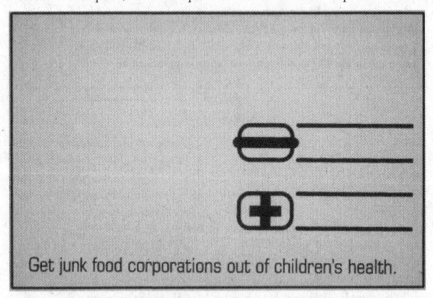

Figure 8.3 Protest postcard distributed to users of the Starship children's hospital on announcement that a McDonald's would be opening.

Figure 8.4 Advertisement for *Middlemore*, a New Zealand TV reality show featuring unrehearsed medical encounters in a public emergency room. Source: *New Zealand Listener*.

too. In Cuba, for instance, the state has been involved in promoting itself as a destination for 'health tourism', in terms of both its therapeutic landscapes and the availability there of high-quality, but cheap, medical care (Feinsilver 1993). Our argument that consumption is pervasive seems particularly relevant where new forms of health care delivery are establishing themselves both in the built environment and through other means. New technologies are playing a part in new forms of delivery, gaining a foothold with medical websites adding to health care 'info-tainment' shows on television as sites of exposure to information. In exposing the viewer to the drama of medical mishap, 'reality TV' is also shaping consumer expectations (see Figure 8.4). In so doing, entrepreneurs are attempting to alter the image of medical care and the type of product to be consumed. To understand such changes requires an appreciation not only of the images and why they are produced, but also of how the images are perceived by patients. Exploring these images adds to our understanding of the meanings of the landscapes of health care as well as the implications of the culture of consumerism in the health care sector.

Further reading

Our placement of this 'case study' chapter at the end of a series of thematic chapters has been to signal our view that it is a promising area for further investigation. Thus for further reading, we reiterate the importance of the few studies that have explicitly begun the process of developing a geography of health care consumption (Barnett and Kearns 1996; Geores 1998, Kearns and Barnett 1992, 1997, 1999). Two further books warrant consideration:

Joseph, A. E. and Phillips, D. R. (1984) *Accessibility and Utilization: Geographical Perspectives on Health Care Delivery*, New York: Harper and Row. This very useful book outlines ways of examining utilization behaviour (an earlier way of expressing consumption) from a largely positivist perspective. We believe that familiarity with such approaches provides a useful foundation for departures using new cultural perspectives. Another strength of this book is its (albeit dated) survey of health care delivery systems, a theme beyond the brief of the present book.

Bell, D. and Valentine, G. (1997) *Consuming Geographies: We Are Where We Eat*, London: Routledge. This book is concerned with food (a topic not unrelated to health!) and can be read as a springboard for those looking for themes to develop and apply within the domains of health and health care.

9

CONCLUSION

Introduction

Health geography is a dynamic field of study. It has witnessed dramatic changes over the past few decades, moving from its original solid foundation in positivist thought through a variety of alternative theoretical and methodological approaches. Along the way, it has tackled many of the most significant problems that face the efficient, equitable and caring delivery of health care. We hope this text has captured some of the excitement that health geographers feel when they attempt to improve the health of people worldwide. In particular, we have tried to show how culture and place interact in producing healthy or unhealthy societies. This chapter summarizes some of the highlights of the culture, health and place synergy. The focus will be on several topics that we want to leave as a final impression on the minds of readers: (1) theoretical diversity, (2) key questions, (3) the disciplinary/interdisciplinary tension, and (4) what makes a health geographer today.

Theoretical diversity

When geographers first became interested in health care issues in the 1960s, the primary theoretical approach open to them was positivism. Since that time, several more options have appeared, such as structuralism, humanism and poststructuralism. Social scientists, geographers and health geographers have utilized these approaches in a variety of ways. Some have clung fiercely to one philosophical position and condemned all others. Others have tried such blends as structuration or critical realism. Still others have been criticized for seizing on the latest theory to become fashionable, whether it applied to their work or not.

The upshot of this theoretical pluralism is that the would-be health geographer is presented with a bewildering array of theoretical concepts. What particular ideas should inform one's research? The answer to that question, we feel, is that it depends on the research question. That statement seems simple enough. However, matching question to theory is usually a thorny problem. If we ask the broadest type of question, such as 'How can I improve the health care system in this place?', any one (or two or more) of a number of approaches could be followed. One could take a positivist approach and map out hospital beds-to-population ratios for different areas in a study site and suggest that inequalities in the ratios be rectified. A structuralist might investigate the role of underlying social forces such as class, gender and race in determining who gets health care and what quality of care. A focus on people's feelings about their contacts with the health care system would be informed by humanistic ideas. Taking a post-structuralist approach, one might examine how power is exercised within medical situations, or study carefully the language used in practitioner–patient interviews. As was outlined at the end of Chapter 2, each of the substantive chapters (3 to 8) focuses on one (and occasionally two) of the principal theoretical approaches. It should be noted, however, that a research question might be addressed by taking more than one approach. Over time, health geographers appear to have become more willing to employ ideas from different philosophical stances that complement each other.

No matter what health care problem one tackles, it cannot be overemphasized that, first, *some* theoretical development should be present and, second, theoretical ideas should be tied to practice. The first point is important, because results that are not informed by theory have little meaning. Taking up the second point, we find that some students seem quite happy to write up a theoretical section of a thesis or dissertation and then ignore it as they collect, analyse, and discuss their data. Obviously, theory is not much use if treated in this way. An entire research project should consistently refer back to theoretical concepts. Did one's findings conform to theory, contradict received ideas, or move theory along in a new direction? No doubt there is still space for health geographers to write purely theoretical papers, but, for the most part, we envision work that draws upon the exciting theoretical advances made in recent years and applies them to practical problems.

Before we leave this section, something should be said about methods. Here again one encounters pluralism. At the most basic level, there are quantitative and qualitative techniques that are available to health geographers. For the most part, quantitative methods are used by those with a positivist bent, but positivists may be able to use qualitative data and non-positivists may supplement their qualitative research with some measurable items. Clearly, the work that health geographers have been doing lately calls for mainly qualitative assessments and that is why Chapter 3 was devoted to qualitative methodologies. Such techniques as in-depth interviews and participant observation are essential to the examination of the trilogy of culture, health and place. These methods, contrary to the belief of some, can (and certainly should) be used in a rigorous and careful manner. The insights they provide are what drives the devoted health geographer.

Key questions

What kinds of questions motivate the health geographer? Most of us find that the delivery of health care is far from adequate, even in the economically most well-off countries, and we want to know why, and what might be done about it. The goals of a health geographer would be similar to those of anyone who wants to improve health care systems: equality for all, affordability, efficiency, accessibility, caring attitudes from health practitioners. The geographer tries to satisfy these goals using concepts current in the discipline. In the early days of health geography, there was a heavy reliance on spatial analysis to investigate questions of resource distributions and physical access to care. There has, however, been a turn to more *placial* analysis, investigating what internal and external influences there are on how people receive health care in specific settings. The major thrust of this text has been on how culture comes into the health and place mix; this defines for us some key questions. In what follows, we set out a series of queries, not in terms of specific research projects (we hope this is amply done within the text), but in terms of broad sets of questions that are informed by social/cultural theories. These questions both summarize much of the text and are intended to stimulate the thinking of students. They are not meant to be exhaustive by any means. Neither are they specific research questions. Rather, they are suggestive of the kinds of queries that we feel health geographers might be thinking about.

Questions involving the importance of culture (Chapter 2)

How do various aspects of culture such as politics, religion and eating habits affect health?
How is health care regionalized, diffused, and related to the natural environment?
How can theoretical approaches inform the work of health geographers?

Questions about conducting research (Chapter 3)

How does one observe cultures of health in places?
What qualitative techniques can be used in health care research?
How does one place themselves within a research setting?

Structural questions (Chapter 4)

How do underlying social forces such as class, gender and ethnicity affect health in places?

How can health care studies be made sociospatially relevant?

Why are some places unhealthy?

What are the effects of political systems and economic changes (e.g. restructuring) on health in places?

How do dominant ideologies such as medicalization affect care and how can they be resisted?

Agency questions (Chapter 4)

What beliefs do people living in a place have about their health and health care provision?

How do individuals feel about their interactions with health care systems?

How do cultural symbols play a role in health care perceptions?

Why are some people deprived of basic health care?

Questions blending structure and agency (Chapter 4)

How do underlying social forces and human actions interact to create healthy or unhealthy places?

How can time geography be used to help study health care?

Questions based on the role of language (Chapter 5)

How does the language used in medical settings affect the nature and quality of health care?

In what ways does naming become norming in places?

How do metaphors, semantic networks and narratives create meaning in places?

How can we trace knowledge and power relations through health care situations?

In what ways do various media mediate people's perceptions about health care?

Questions that deal with difference (Chapter 6)

How can poststructuralist/postmodern ideas about difference and deviance, the body and 'the other' inform studies of health in places?

How can the health of women be improved using concepts from feminist geography?

How do the health care experiences of people from different ethnic backgrounds compare?

How are disease and health related to geographies of sexuality?

How well do people with disabilities cope with their environments at different spatial scales?

Healing landscape questions (Chapter 7)

What role does nature play in the healing process?

In what ways are therapeutic landscapes social constructs?

How do human perceptions as well as concrete and abstract symbols play roles in creating healing landscapes?

How can concepts about therapeutic landscapes be used to study past and present health situations?

Consumerism in health care questions (Chapter 8)

How are symbols and symbolic landscapes used to sell health care?

How do architecture and advertising influence people's health care behaviour in places?

How do health care organizations market specific health care products?

The disciplinary/interdisciplinary tension

The reader of this text should have become aware that the health geographer is both steeped in the discipline of geography and interdisciplinary. There is a useful tension between having a solid foundation in one's own field and at the same time being open to influences from the other social sciences, the humanities, the natural sciences and the medical sciences. We take issue with those geographers who insist on the purity of their discipline and are afraid of straying beyond its boundaries. Such a stance makes little sense, because disciplinary boundaries are fuzzy, permeable and continually changing.

A great deal of the excitement that has been generated in geography as a whole and health geography in particular over the past couple of decades stems from this tension. On the one hand, we have borrowed many concepts from colleagues in other disciplines, such as explanatory models from anthropologists, underlying social forces from sociologists, and world systems theory from political scientists. Meanwhile, many scholars in other disciplines have been quite happy to use the language and concepts of geography in their theoretical formulations and practice. Thus, as examples, the French theorist Michel Foucault writes in *The Birth of the Clinic*, 'This book is about space' (Foucault: 1976, ix) and the sociologists Macintyre, Maciver and Soomans (1993) emphasize the importance of place to health.

The interdisciplinary/disciplinary tension encourages the health geographer to move in two directions at the same time. First, he or she should be *disciplined*; that is, should be immersed in placial and spatial thinking. Concurrently, one should be *undisciplined*, open to the myriad of ideas that are available outside geography. A good example of this duality would be the theoretical landscape or healthy place idea. Obviously, place and landscape are geographic concepts that are layered with disciplinary meanings. But the idea also borrows from history, ecology, political science, anthropology, sociology and linguistics. Of course, what one blends from geography and other disciplines has to make sense, tell a coherent story. In a poststructuralist world, there is a danger of becoming fragmented and confused, but this is countered by the energizing possibilities for creating new syntheses, new intellectual identities.

What makes a health geographer today?

Throughout this text we have tried to show that being a placial, cultural, health geographer is an exciting and rewarding career. We have also thought about some of the qualities that such a hybrid might possess. To begin with, of course, one should be deeply interested in the intimate details of how culture and health interact in specific settings. The health geographer is interested in people's ways of life, their ways of thinking and acting. This means a willingness to engage in intensive fieldwork, using such techniques as in-depth interviews and participant observation. At the same time, one must understand the observer's place within the research setting. In the field, one holds lengthy conversations with people, learns to listen to them carefully, and also has all one's senses attuned to the surrounding physical and human-made environments.

The health geographer is theoretically knowledgeable, up to date on the social and cultural debates that inform her or his work. At the same time, he or she is interested in the practical results of research; one is an activist, an advocate for the marginalized and underserved. Health geography studies should combine methodological rigour with sympathy for the researched, often a fine line to tread. Finally, the health geographer is alert to aspects of health care delivery that one ordinarily misses, such as people's feelings and experiences, concrete and abstract symbols in the landscape, human diversity, inclusion and exclusion, and the stories that people tell.

BIBLIOGRAPHY

Abel, S. and Kearns, R. A. (1991) 'Birth places: a geographical perspective on planned home birth in New Zealand', *Social Science and Medicine* 33, 7: 825–834.

Adams, P. C. (1992) 'Television as gathering place', *Annals of the Association of American Geographers* 82, 1: 117–135.

Ai, A. L., Dunkle, R. E. and Bolling, S. F. (1998) 'The role of private prayer in psychological recovery among mid-life and aged patients following cardiac surgery', *The Gerontologist* 38, 5: 591–601.

Aitken, S. C. and Zonn, L. E. (1994) 'Re-presenting the place pastiche', in S. C. Aitken and L. E. Zonn (eds) *Place, Power, Situation, and Spectacle: A Geography of Film*, Totowa, NJ: Rowman and Littlefield.

Alderman, D. H. (1997) 'TV news hyper-coverage and the representation of place: observations on the O. J. Simpson case', *Geografiska Annaler* 79B: 83–95.

Andes, N. (1989) 'Socioeconomic, medical care, and public health contexts affecting infant mortality: a study of community-level differentials in Peru', *Journal of Health and Social Behavior* 3: 386–397.

Anspach, R. R. (1988) 'Notes on the sociology of medical discourse: the language of case presentation', *Journal of Health and Social Behavior* 29: 357–375.

Arcury, T., Gesler, W. M. and Cook, H. L. (1999) 'Meaning in the use of unconventional arthritis therapies', *American Journal for Health Prevention* 14, 1: 7–15.

Arcury, T. A., Quandt, S. A., McDonald, J. and Bell, R. A. (2000) 'Faith and health self-management of rural older adults', *Journal of Cross-Cultural Gerontology* 15: 55–74.

Ascot Integrated Hospital (1999) *Taking Health Care into the Future*, Auckland: AIH.

Ashton, T. (1992) 'Reform of the health services: weighing up the costs and benefits', in J. Boston and P. Dalziel (eds) *The Decent Society: Essays on National's Social Policy*, Auckland: Auckland University Press.

Asthana, S. (1996) 'Women's health and women's empowerment: a locality perspective', *Health and Place* 2, 1: 1–13.

Atkinson, P. (1995) *Medical Talk and Medical Work: The Liturgy of the Clinic*, London: Sage.

Ayensu, E. S. (1981) 'A worldwide role for the healing powers of plants', *The Smithsonian* 12: 86–97.

Baer, L. D. (2001) 'The acceptability of international medical graduates in rural America', unpublished Ph.D. dissertation, Department of Geography, University of North Carolina, Chapel Hill, NC.

Baer, L. D. and Good, C. M. Jr (1998) 'The power of the state', in R. J. Gordon, B. C. Nienstedt and W. M. Gesler (eds) *Alternative Therapies: Expanding Options in Health Care,* New York: Springer.

Baer, L. D., Gesler, W. M. and Konrad, T. R. (2000) 'The wineglass model: tracking the histories of health professionals', *Social Science and Medicine* 50, 6: 317–329.

Baggott, B. (1994) *Health and Health Care in Britain,* New York: St Martin's Press.

Baker, S. R. (1979) 'The diffusion of high technology medical innovation: the computed tomography scanner example', *Social Science and Medicine* 13D: 155–162.

Bamborough, J. B. (1980) *The Little World of Man,* London: Longmans Green.

Barnes, T. J. (1992) 'Reading the texts of theoretical economic geography: the role of physical and biological metaphors', in T. J. Barnes and J. S. Duncan (eds) *Writing Worlds: Discourse, Text and Metaphor in the Representation of Landscape,* London: Routledge.

Barnes, T. J. and Gregory, D. (1997) 'Geography and difference', in T. J. Barnes and D. Gregory (eds) *Reading Human Geography: The Poetics and Politics of Inquiry,* New York: John Wiley.

Barnett, J. R. and Kearns, R. A. (1996) 'Shopping around? Consumerism and the use of private accident and emergency clinics in Auckland', *Environment and Planning A* 28: 1053–1075.

Bashshur, R. L., Shannon, G. W. and Metzner, C. A. (1971) 'Some ecological differentials in the use of medical services', *Health Services Research* 6: 61–75.

Bastien, J. W. (1985) 'Qollohuaya–Andean body concepts: a topographical–hydraulic model of physiology', *American Anthropologist* 87: 595–611.

Bauer, A. (1971) *Thomas Mann,* New York: Frederick Ungar.

Bentham, G., Haynes, R. and Lovett, A. (1991) 'Introduction', *Social Science and Medicine* 33: ix–x.

Berg, L. D. (1994) 'Masculinity, place and a binary discourse of "theory" and "empirical investigation" in the human geography of Aotearoa, New Zealand', *Gender, Place and Culture* 1, 2: 245–260.

Berg, L. D. and Kearns, R. A. (1996) 'Naming as norming: "race", gender, and the identity politics of naming places in Aotearoa/New Zealand', *Environment and Planning D: Society and Space* 14: 99–122.

Berg, L. D. and Kearns, R. A. (1997) 'Constructing cultural geographies of Aotearoa', *New Zealand Geographer* 53, 2: 1–2.

Best, S. and Kellner, D. (1991) *Postmodern Theory: Critical Interrogations,* Basingstoke: Macmillan.

Bhardwaj, S. (1980) 'Medical pluralism and homeopathy: a geographic perspective', *Social Science and Medicine* 14B: 209–216.

Blumhagen, D. (1979) 'The doctor's white coat: the image of the physician in modern America', *Annals of Internal Medicine* 91: 111–116.

Blunt, A. and Wills, J. (2000) *Dissident Geographies: An Introduction to Radical Ideas and Practice,* Harlow: Prentice-Hall.

Boyne, R. (1990) *Foucault and Derrida: The Other Side of Reason*, London: Routledge.

Brookfield, H. C. (1964) 'Questions on the human frontiers of geography', *Economic Geography* 40, 4: 283–303.

Brown, D. M. (1984) 'A peaceable kingdom', *National Geographic Traveler* 1, 2: 103–115.

Brown, M. (1995) 'Ironies of distance: an ongoing critique of the geographies of AIDS', *Environment and Planning D: Society and Space* 13: 159–183.

Bunge, W. (1978) 'The first years of the Detroit Geographical Expedition: a personal report', in R. Peet (ed.) *Radical Geography: Alternative Viewpoints on Contemporary Social Issues*, London: Methuen.

Bury, M. (1998) 'Postmodernity and health', in G. Scambler and P. Higgs (eds) *Postmodernity and Health*, London: Routledge.

Butler, J. (1990) *Gender Trouble: Feminism and the Subversion of Identity*, New York: Routledge.

Buttimer, A. (1979) 'Reason, rationality and human creativity', *Geografiska Annaler* 61B: 43–49.

Buttimer, A. (1982) 'Musing on Helicon: root metaphors and geography', *Geografiska Annaler* 64B: 89–96.

Cain, C. (1991) 'Personal stories: identity acquisition and self-understanding in Alcoholics Anonymous', *Ethos* 19: 210–253.

Cayleff, S. E. (1988) 'Gender, ideology, and the water-cure movement', in N. Gevitz (ed.) *Other Healers: Unorthodox Medicine in America*, Baltimore: Johns Hopkins University Press.

Chiotti, Q. P. and Joseph, A. E. (1995) 'Casey house: interpreting the location of a Toronto AIDS hospice', *Social Science and Medicine* 41, 1: 131–140.

Chouinard, V. (1997) 'Making space for disabling differences: challenging ableist geographies', *Environment and Planning D: Society and Space* 15: 379–390.

Cilliers, P. (1998) *Complexity and Postmodernism: Understanding Complex Systems*, London: Routledge.

Clifford, J. (1988) *The Predicament of Culture*, Cambridge, MA: Harvard University Press.

Cloke, P., Philo, C. and Sadler, D. (1991) *Approaching Human Geography: An Introduction to Contemporary Theoretical Debates*, New York: Guilford Press.

Cohen, M., Tripp-Reimer, T., Smith, C., Sorofman, B. and Lively, S. (1994) 'Explanatory models of diabetes: patient practitioner variation', *Social Science and Medicine* 38, 1: 59–66.

Cordes, S. M. (1988) 'The changing rural environment and the relationship between health services and rural development', *Health Service Research* 23: 757–784.

Cornish, C. V. (1997) 'Behind the crumbling walls; the re-working of a former asylum's geography', *Health and Place* 3, 2: 101–110.

Cornwell, J. (1984) *Hard-Earned Lives*, London: Tavistock.

Cosgrove, D. (2000) 'Iconography', in R. J. Johnston, D. Gregory, G. Pratt and M. Watts (eds) *The Dictionary of Human Geography*, 4th edition, Oxford: Blackwell.

Cosgrove, D. and Daniels, S. (eds) (1988) *The Iconography of Landscape*, Cambridge: Cambridge University Press.

Cosgrove, D. and Jackson, P. (1987) 'New directions in cultural geography', *Area* 19: 95–101.

Craddock, S. (2000) *City of Plagues: Disease, Poverty, and Deviance in San Francisco*, Minneapolis: University of Minnesota Press.

Crang, P. (1996) 'It's showtime: on the workplace geographies of display in a restaurant in southwest England', *Environment and Planning D: Society and Space* 12: 675–704.

Cravey, A., Gesler, W. M., Skelly, A. H., Arcury, T. A. and Washburn, S. A. (2001) 'The role of socio-spatial knowledge networks in chronic disease interventions in communities', *Social Science and Medicine* 52: 1763–1774.

Cresswell, T. (1996) *In Place/Out of Place: Geography, Ideology, and Transgression*, Minneapolis: University of Minnesota Press.

Croft, S. and Beresford, P. (1992) 'The politics of participation', *Critical Social Policy* 35: 20–44.

Csordas, T. J. (1983) 'The rhetoric of transformation in ritual healing', *Culture, Medicine and Psychiatry* 7: 333–375.

Cunliffe, B. (1986) *The City of Bath*, Gloucester: Alan Sutton.

Curtis, S. and Taket, A. (1996) *Health and Societies: Changing Perspectives*, London: Arnold.

Curtis, S., Gesler, W. M., Smith, G. and Washburn, S. A. (2000) 'Approaches to sampling and case selection in qualitative research: examples in the geography of health', *Social Science and Medicine* 50, 7/8: 1001–1014.

Cutchin, M. P. (1997) 'Physician retention in rural communities: the perspective of experiential place integration', *Health and Place* 3, 1: 25–41.

Day, P. and Klein, R. (1991) 'Britain's health care experiment', *Health Affairs* 10: 39–59.

Dear, M. and Wolch, J. R. (1987) *Landscapes of Despair: From Deinstitutionalization to Homelessness*, Princeton, NJ: Princeton University Press.

DeBruyn, M. (1992) 'Women and AIDS in developing countries', *Social Science and Medicine* 34, 3: 249–262.

DeSouza, A. and Porter, P. (1974) *The Underdevelopment and Modernization of the Third World*, Washington, DC: Association of American Geographers, Commission on College Geography.

DiGiacomo, S. M. (1987) 'Biomedicine as a cultural system: an anthropologist in the kingdom of the sick', in H. A. Baer (ed.) *Encounters with Biomedicine: Case Studies in Medical Anthropology*, New York: Gordon and Breach Science Publishers.

Dobbs, G. R. (1997a) 'Interpreting the Navajo sacred geography as a landscape of healing', unpublished MA thesis, Department of Geography, University of North Carolina, Chapel Hill, NC.

Dobbs, G. R. (1997b) 'Interpreting the Navajo sacred geography as a landscape of healing', *Pennsylvania Geographer* 35, 2: 136–150.

Dobbs, G. R. (2001) 'The urban experience of disability', unpublished manuscript.

Dorn, M. and Laws, G. (1994) 'Social theory, body politics, and medical geography: extending Kearns's invitation', *The Professional Geographer* 46, 1: 106–110.

Dow, J. (1986) 'Universal aspects of symbolic healing: a theoretical synthesis', *American Anthropologist* 88: 56–69.

Dubos, R. (1959) *The Mirage of Health: Utopian Progress and Biological Change*, New York: Ander Books.

Duggan, L. (1986) 'From birth control to population control: Depo-Provera in Southeast Asia', in K. McDonnell (ed.) *Adverse Effects: Women and the Pharmaceutical Industry*, Toronto: Women's Educational Press.

Duncan, J. (1980) 'The superorganic in American cultural geography', *Annals of the Association of American Geographers* 70: 181–198.

Duncan, J. (1993) 'Sites of representation: place, time and discourse of the Other', in J. Duncan and D. Ley (eds) *Place/Culture/Representation*, London: Routledge.

Duncan, J. (2000) 'Place', in R. J. Johnston, D. Gregory, G. Pratt and M. Watts (eds) *The Dictionary of Human Geography*, 4th edition, Oxford: Blackwell.

Duncan, J. and Duncan, N. (1988) '(Re)reading the landscape', *Environment and Planning D: Society and Space* 6: 117–126.

Durie, M. (1994) *Whaiora: Maori Health Development*, Auckland: Oxford University Press.

Dyck, I. (1998) 'Women with disabilities and everyday geographies', in R. A. Kearns and W. M. Gesler (eds) *Putting Health into Place: Landscape, Identity, and Well-being*, Syracuse, NY: Syracuse University Press.

Dyck, I. (1999) 'Using qualitative methods in medical geography: deconstructive moments in a subdiscipline?', *The Professional Geographer* 51, 2: 243–253.

Dyck, I. and Kearns, R. (1995) 'Transforming the relations of research: towards culturally safe geographies of health and healing', *Health and Place* 1, 3: 137–147.

Eisenberg, D. M., Kessler, R. C., Foster, C., Norlock, F. E., Calkins, D. R. and Delbanco, T. L. (1993) 'Unconventional medicine in the United States: prevalence, costs and patterns of use', *New England Journal of Medicine* 328: 246–252.

Elliott, S. J. (1998) 'Why are some women healthy and others not?', Paper presented at the Association of American Geographers Annual Meeting, March 1998, Boston.

Elliott, S. J. (1999) 'And the Question shall determine the Method', *Professional Geographer* 51: 240–243.

Emerson, J. P. (1970) 'Behavior in private places: sustaining definitions of reality in gynecological examinations', in H. P. Dreitzel (ed.) *Patterns of Communication Behavior*, New York: Macmillan.

Entrikin, J. N. (1991) *The Betweenness of Place: Towards a Geography of Modernity*, Baltimore: Johns Hopkins University.

Epp, J. (1986) *Achieving Health for All: A Framework for Health Promotion*, Ottawa: Ministry of Supply and Services.

Estroff, S. (1981) *Making It Crazy: An Ethnography of Psychiatric Clients in an American Community*, Los Angeles: University of California Press.

Evans, M. (1988) 'Participant observation: the researcher as research tool', in J. Eyles and D. M. Smith (eds) *Participant Observation: The Researcher as Research Tool*, London: Polity Press.

Evans, R. G., Barer, M. L. and Marmor, T. R. (1994) *Why Are Some People Healthy and Others Not?*, New York: Aldine de Gruyter.

Eyles, J. (1985) *Senses of Place*, Warrington: Silverbrook Press.

Eyles, J. (1987) 'Housing advertisements as signs: locality creation and meaning-systems', *Geografiska Annaler* 69B: 93–105.

Eyles, J. and Woods, K. (1982) *The Social Geography of Health and Health Care*, London: Croom Helm.

Fabrega, H. Jr (1975) 'The need for an ethnomedical science', *Science* 189: 969–975.

Fabrega, H. Jr (1977) 'The scope of ethnomedical science', *Culture, Medicine and Psychiatry* 1: 201–228.

Fabrega, H. Jr (1980) *Disease and Social Behavior: An Interdisciplinary Perspective*, Cambridge, MA: MIT Press.

Feinsilver, J. M. (1993) *Healing the Masses: Cuban Health Politics at Home and Abroad*, Berkeley: University of California Press.

Fisher, S. (1982) 'The decision-making context: how doctors and patients communicate', in R. J. Di Pietro (ed.) *Lingusitics and the Professions*, Norwood, NJ: Ablex.

Fisher, S. (1991) 'A discourse of the social: medical talk/power talk/oppositional talk?', *Discourse and Society* 2, 2: 157–182.

Foote, K. E., Hugill, P. J. and Mathewson, K. (1994) *Re-reading Cultural Geography*, Austin: University of Texas Press.

Forbes, S. (1992) *The Nature of Denali: Denali National Park*, Alaska: Alaska Natural History Association.

Foucault, M. (1965) *Madness and Civilization: A History of Madness in the Age of Reason*, New York: Random House.

Foucault, M. (1975) *The Birth of the Clinic: An Archaelogy of Medical Perception*, New York: Vintage Books.

Foucault, M. (1976) *The Birth of the Clinic: An Archaeology of Medical Perception*, trans. A. M. Sheridan, London: Tavistock Publications.

Foucault, M. (1977) *Discipline and Punish: The Birth of the Prison*, New York: Pantheon Books.

Foucault, M. (1978) *The History of Sexuality*, vol. 1: *An Introduction*, trans. R. Hurley, New York: Pantheon.

Fox, N. J. (1993a) *Postmodernism, Sociology, and Health*, Buckingham: Open University Press.

Fox, N. J. (1993b) 'Discourse, organisation and the surgical ward round', *Sociology of Health and Illness* 15, 1: 16–42.

Frazier, L. J. and Scarpaci, J. L. (1998) 'Landscapes of state violence and the struggle to reclaim community', in R. A. Kearns and W. M. Gesler (eds) *Putting Health into Place: Landscape, Identity, and Well-Being*, Syracuse, NY: Syracuse University Press.

Game, A. (1991) *Undoing the Social: Towards a Deconstructive Sociology*, Milton Keynes: Open University Press.

Geertz, C. (1973) 'Thick description: toward an interpretive theory of culture', in C. Geertz (ed.) *The Interpretation of Cultures*, New York: Basic Books.

Geores, M. (1998) 'Surviving on metaphor: how "Health = Hot Springs" created and sustained a town', in R. A. Kearns and W. M. Gesler (eds) *Putting Health into Place: Landscape, Identity, and Well-Being*, Syracuse: Syracuse University Press.

Geores, M. and Gesler, W. M. (1999) 'Compromised space: contests over the provision of a therapeutic environment for people with mental illness', in A. Williams (ed.) *Therapeutic Landscapes: The Dynamic between Place and Wellness*, Lanham, MD: University Press of America.

Gesler, W. M. (1989) 'The role of multinational pharmaceutical firms in health care privatization in developing countries', in J. L. Scarpaci (ed.) *Health Services Privatization in Industrial Societies*, New Brunswick, NJ: Rutgers University Press.

Gesler, W. M. (1991) *The Cultural Geography of Health Care*, Pittsburgh: University of Pittsburgh Press.

Gesler, W. M. (1992) 'Therapeutic landscapes: medical geographic research in light of the new cultural geography', *Social Science and Medicine* 34, 7: 735–746.

Gesler, W. M. (1993) 'Therapeutic landscapes: theory and a case study of Epidauros, Greece', *Environment and Planning D: Society and Space* 11: 171–189.

Gesler, W. M. (1994) 'The global pharmaceutical industry: health, development and business', in D. R. Phillips and Y. Verhasselt (eds) *Health and Development*, London: Routledge.

Gesler, W. M. (1996) 'Lourdes: healing in a place of pilgrimage', *Health and Place* 2, 2: 95–105.

Gesler, W. M. (1998) 'Bath's reputation as a healing place', in R. A. Kearns and W. M. Gesler (eds) *Putting Health into Place: Landscape, Identity, and Well-Being*, Syracuse, NY: Syracuse University Press.

Gesler, W. M. (1999) 'Words in wards: language, health and place', *Health and Place* 5, 1: 13–25.

Gesler, W. M. (2000) 'Hans Castorp's journey to knowledge of disease and health in Thomas Mann's *The Magic Mountain*', *Health and Place* 6, 2: 125–134.

Gesler, W. M. and Gordon, R.J. (1998) 'Alternative therapies: why now?', in R. J. Gordon, B. C. Nienstedt and W. M. Gesler (eds) *Alternative Therapies: Expanding Options in Health Care*, New York: Springer.

Gesler, W. M. and Ricketts, T. C. (eds) (1992) *Health in Rural North America: The Geography of Health Care Services and Delivery*, New Brunswick, NJ: Rutgers University Press.

Gesler, W. M., Bird, S. T. and Oljeski, S. A. (1997) 'Disease ecology and a reformist alternative: the case of infant mortality', *Social Science and Medicine* 44, 5: 657–671.

Gesler, W. M., Blundell, B. and Spence, M. (1998) 'Creating therapeutic environments in hospitals: literature review and case study', unpublished manuscript.

Gesler, W. M., Arcury, T. A. and Koenig, H. G. (2000a) 'The effects of religion on health among rural older adults with different cultural backgrounds', *Journal of Crosscultural Gerontology* 15:1–2.

Gesler, W. M., Arcury, T. A. and Cook, H. (2000b) 'Meanings given to the causes of osteoarthritis by residents of a rural North Carolina county', unpublished manuscript.

Giddens, A. (1976) *New Rules of Sociological Method*, New York: Basic Books.

Gillespie, G. (1998) ' Havens for the fashionable and sickly: society, sickness and space at nineteenth century southern spring resorts', unpublished Ph.D. dissertation, Department of Geography, University of North Carolina, Chapel Hill.

Glacken, C. J. (1967) *Traces on the Rhodian Shore*, Berkeley: University of California Press.

Gleeson, B. (1999) *Geographies of Disability*, London: Routledge.

Gleeson, B. and Kearns, R. (2001) 'Re-moralising landscapes of care', *Environment and Planning D: Society and Space* 19, 1: 61–80.

Glennie, P. D. and Thrift, N. J. (1992) 'Modernity, urbanism, and modern consumption', *Environment and Planning D: Society and Space* 10: 423–443.

Godkin, M. A. (1980) 'Identity and place: clinical applications based on notions of rootedness and uprootedness', in A. Buttimer and D. Seamon (eds) *The Human Experience of Space and Place*, London: Croom Helm.

Goffman, E. (1963) *The Presentation of the Self in Everyday Life*, Harmondsworth: Penguin.

Gold, M. (1985) 'A history of nature', in D. Massey and J. Allen (eds) *Geography Matters!*, Cambridge: Cambridge University Press.

Golledge, R. G. (1993) 'Geography and the disabled: a survey with special reference to vision impaired and blind populations', *Transactions of the Institute of British Geographers* NS 18, 1: 63–85.

Good, B. J. (1977) 'The heart of what's the matter: the semantics of illness in Iran', *Culture, Medicine and Psychiatry* 1: 25–58.

Good, B. J. (1994) *Medicine, Rationality, and Experience: An Anthropological Perspective*, Cambridge: Cambridge University Press.

Good, C. M. (1987) *Ethnomedical Systems in Africa: Patterns of Traditional Medicine in Rural and Urban Keyna*, New York: Guilford Press.

Gordon, D. R. (1990) 'Embodying illness, embodying cancer', *Culture, Medicine, and Psychiatry* 14: 275–297.

Gordon, R. J. and Silverstein, G. (1998) 'Marketing channels for alternative health care', in R. J. Gordon, B. C. Nienstedt and W. M. Gesler (eds) *Alternative Therapies: Expanding Options in Health Care*, New York: Springer.

Goss, J. (1993) 'The "magic of the mall": an analysis of form, function, and meaning in the contemporary retail built environment', *Annals of the Association of American Geographers* 83, 1: 18–47.

Gould, S. J. (1990) 'Taxonomy as politics: the harm of false classification', *Dissent* (Winter): 73–78.

Gramsci, A. (1971) *Selections from the Prison Notebooks*, ed. and trans. Q. Hoare and G. Nowell-Smith, London: Lawrence and Wishart.

Gregory, D. (1981) 'Human agency and human geography', *Transactions of the Institute of British Geographers*, NS. 6: 1–18.

Gregory, D. (2000a) 'Positivism', in R. J. Johnston, D. Gregory, G. Pratt and M. Watts (eds) *The Dictionary of Human Geography*, 4th edition, Oxford: Blackwell.

Gregory, D. (2000b) 'Structuralism', in R. J. Johnston, D. Gregory, G. Pratt and M. Watts (eds) *The Dictionary of Human Geography*, 4th edition. Oxford: Blackwell.

Gregson, N. (1995) 'And now it's all consumption?', *Progress in Human Geography* 19: 135–141.

Gupta, A. and Ferguson, J. (1997) *Anthropological Locations: Boundaries and Grounds of a Field Science*, Berkeley: University of California Press.

Hagerstrand, T. (1982) 'Diorama, path, and project', *Tijdschrift voor Economische en Sociale Geografie* 73: 323–329.

Hagey, R. (1984) 'The phenomenon, the explanations and the responses: metaphors surrounding diabetes in urban Canadian Indians', *Social Science and Medicine* 18: 265–272.

Haggett, P. (1975) *Geography: A Modern Synthesis*, New York: Harper and Row.

Hall, E. (2000) ' "Blood, brain and bones": taking the body seriously in the geography of health and impairment', *Area* 32: 21–29.

Hall, E. and Kearns, R. (2001) 'Making space for the 'intellectual' in geographies of disability', *Health and Place* 7, 237–46.

Hanson, S. and Pratt, G. (1995) *Gender, Work and Space*, New York: Routledge.

Haraway, D. J. (1991) *Simians, Cyborgs, and Women: The Reinvention of Nature*, London: Free Association Books.

Hart, A. C., Schmidt, K. M. and Aaron, W. S. (eds) (1989) *St. Anthony's ICD.9.CM Code Book*, Reston, VA: St Anthony Publishing.

Harvey, D. (1973) *Social Justice and the City*, Baltimore: Johns Hopkins University Press.

Harvey, D. (1989) *The Condition of Postmodernity*, Oxford: Blackwell.

Harvey, D. (1996) *Justice, Nature and the Geography of Difference*, Oxford: Blackwell.

Hayes, M. V. (1999) 'Population health promotion: responsible sharing of future directions', *Canadian Journal of Public Health* (Nov./Dec.): 15–17.

Helman, C. G. (1994) *Culture, Health and Illness*, 3rd edition, Oxford: Butterworth Heinemann.

Hudson, B. (1977) 'The new geography and the new imperialism: 1870–1918', *Antipode* 9, 2: 12–19.

Hughes, C. C. and Hunter, J. M. (1970) 'Disease and "development" in tropical Africa', *Social Science and Medicine* 3: 443–493.

Hunt, L. M., Arar, N. H. and Larme, A. C. (1998) 'Contrasting patient and practitioner perspectives in NIDDM management', *Western Journal of Nursing Research* 20, 6: 656–682.

Hunter, J. M. (1973) 'Geophagy in Africa and the United States: a culture–nutrition hypothesis', *Geographical Review* 63: 170–195.

Hunter, J. M., Horst, O. H. and Thomas, R. N. (1989) 'Religious geophagy as a cottage industry: the holy clay tablets of Esquipulas, Guatemala', *National Geographic Research* 5, 3: 281–295.

Hutton, J. D. and Richardson, L. D. (1995) 'Healthscapes: the role of facility and physical environment on consumer attitudes, satisfaction, quality assessments, and behaviors', *Health Care Management Review* 20, 2: 48–61.

Illich, I. (1976) *Medical Nemesis: The Expropriation of Health*, New York: Pantheon Books.

Isajiw, W. (1974) 'Definitions of ethnicity', *Ethnicity* 1: 111–124.

Ityavyar, D. A. (1987) 'Background to the development of health services in Nigeria', *Social Science and Medicine* 24, 6: 487–499.

Jackson, P. (1983) 'Principles and problems of participant observation', *Geografiska Annaler* 65B: 39–46.

Jackson, P. (1989) *Maps of Meaning: An Introduction to Cultural Geography*, London: Unwin Hyman.

Jackson, P. (1993) 'Changing ourselves: a geography of position', in R. J. Johnston (ed.) *Changing Ourselves: A Geography of Position*, Oxford: Basil Blackwell.

Jackson, P. (2000) 'other/otherness', in R. J. Johnston, D. Gregory, G. Pratt and M. Watts (eds) *The Dictionary of Human Geography*, Oxford: Blackwell.

Jackson, P. and Penrose, J. (eds) (1993) *Constructions of Race, Place and Nation*, London: UCL Press.

Jackson, P. and Smith, S. J. (1984) *Exploring Social Geography*, London: George Allen and Unwin.

Jackson, P. and Taylor, J. (1996) 'Geography and the cultural politics of advertising', *Progress in Human Geography* 20: 256–371.

Jackson, P. and Thrift, N. (1995) 'Geographies of consumption', in D. Miller, (ed.) *Acknowledging Consumption*, London: Routledge.

James, P. E. (1972) *All Possible Worlds: A History of Geographical Ideas*, New York: Bobbs-Merrill.

Jeffrey, R. (1979) 'Normal rubbish: deviant patients in casualty departments', *Sociology of Health and Illness* 1, 1: 90–107.

Jefkins, F. (1994) *Advertising*, London: Pitman.

Johnston, R. J. (1986a) *On Human Geography*, Oxford: Basil Blackwell.

Johnston, R. J. (1986b) *Philosophy and Human Geography: An Introduction to Contemporary Approaches*, London: Edward Arnold.

Johnston, R. J., Gregory, D., Pratt, G. and Watts, M. (eds) (2000) *The Dictionary of Human Geography*, 4th edition, Oxford: Blackwell.

Jones, K. and Moon, G. (1987) *Health, Disease and Society: An Introduction to Medical Geography*, London: Routledge and Kegan Paul.

Joseph, A. E. and Phillips, D. R. (1984) *Accessibility and Utilization: Geographic Perspectives on Health Care Delivery*, New York: Harper and Row.

Katz, C. (1994) 'Playing the field: questions of fieldwork in geography', *The Professional Geographer* 46: 67–72.

Katz, P. (1981) 'Ritual in the operating room', *Ethnology* 20, 4: 335–350.

Kearns, R. A. (1987) 'In the shadow of illness: a social geography of the chronically mentally disabled in Hamilton, Ontario', Ph.D. dissertation, McMaster University.

Kearns, R. A. (1991) 'The place of health in the health of place: the case of the Hokianga special medical area', *Social Science and Medicine* 33: 519–530.

Kearns, R. A. (1993) 'Place and health: towards a reformed medical geography', *The Professional Geographer* 45: 139–147.

Kearns, R. A. (1995) 'Medical geography: making space for difference', *Progress in Human Geography* 19, 2: 249–257.

Kearns, R. A. (1996) 'AIDS and medical geography: embracing the other?', *Progress in Human Geography* 20, 1: 123–131.

Kearns, R. A. (1997a) 'Constructing (bi)cultural geographies: research on, and with, people of the Hokianga District', *New Zealand Geographer* 52: 3–8.

Kearns, R. A. (1997b) 'Consumerist ideology and the symbolic landscapes of private medicine', *Health and Place* 3, 3: 171–180.

Kearns, R. A. (1997c) 'Narrative and metaphor in health geographies', *Progress in Human Geography* 21: 269–277.

Kearns, R. A. and Barnett, J. R. (2000) 'Happy meals in the Starship *Enterprise*: interpreting a moral geography of food and health care consumption', *Health and Place* 6, 2: 81–93.

Kearns, R. A. (2000) 'Being there: research through observing and participating', in I. Hay (ed.) *Qualitative Methods in Human Geography*, Melbourne: Oxford, University Press.

Kearns, R. A. and Barnett, J. R. (1992) 'Enter the supermarket: entrepreneurial medical practice in New Zealand', *Environment and Planning C: Government and Policy* 10: 267–281.

Kearns, R. A. and Barnett, J. R. (1997) 'Consumerist ideology and the symbolic landscapes of private medicine', *Health and Place* 3, 3: 171–180.

Kearns, R. A. and Barnett, J. R. (1999) 'To boldly go? Place, metaphor and marketing of Auckland's Starship hospital', *Environment and Planning D: Society and Space* 17: 201–226.

Kearns, R. A. and Collins, C. A. (2000) 'New Zealand children's health camps: therapeutic landscapes meet the contract state', *Social Science and Medicine* 51: 1047–1059.

Kearns, R. A. and Gesler, W. M. (1998) *Putting Health into Place: Landscape, Identity and Well-Being*, Syracuse, NY: Syracuse University Press.

Kearns, R. A. and Joseph, A. E. (2000) 'Contracting opportunities: intepreting post-asylum geographies of mental health care in Auckland, New Zealand', *Health and Place* 6: 159–170.

Kearns, R. and Moon, G. (2001) 'Abstracting health from medical geography: novelty, place and theory after a decade of change', unpublished manuscript (available from the second author).

Kearns, R. A. and Newman, D. (1999) 'The power of discourse and monument: Ascot Hospital's place in millennial Auckland', *Proceedings, New Zealand Geographical Society Conference*, Hamilton: New Zealand Geographical Society.

Kearns, R. A., Murphy, L. and Freison, W. (2001) 'Shopping!', in C. Bell, (ed.) *Everyday Sociologies*, Palmerston North: Dunmore Press.

Kelly, M. P., Davies, J. K. and Charlton, B. G. (1993) 'Healthy cities: a modern problem of a postmodern solution?', in M. P. Kelly and J. K. Davis (eds) *Healthy Cities: A Modern Problem of a Postmodern Solution?*, London: Routledge.

Kenny, C. and Canter, D. (1979) 'Evaluating acute general hospitals', in D. Canter and S. Canter (eds) *Designing for Therapeutic Environments*, New York: John Wiley.

Kirmayer, L. J. (1988) 'Mind and body as metaphors: hidden values in biomedicine', in M. Lock and D. Gordon (eds) *Biomedicine Examined*, Dordrecht: Kluwer Academic Publishers.

Kivell, L. (1995) 'Sex/gender and "race": constructing a harvest workforce', MA thesis, Department of Geography, University of Auckland.

Kleinman, A. M. (1973) 'Medicine's symbolic reality: on a central problem in the philosophy of medicine', *Inquiry* 16: 206–213.

Kleinman, A. M. (1978) 'Concepts and a model of comparison of medical systems as cultural systems', *Social Science and Medicine* 12: 85–93.

Kleinman, A. M. (1980) 'Major conceptual and research issues for cultural (anthropological) psychiatry', *Culture, Medicine, and Psychiatry* 4, 1: 3–13.

Kleinman, A. M. (1988) *The Illness Narratives: Suffering, Healing and the Human Condition*, New York: Basic Books.

Knopp, L. (1995) Sexuality and urban space: a framework for analysis, in D. Bell and G. Valentine (eds) *Mapping Desire: Geographies of Sexualities*, London: Routledge.

Knowles, C. (2000) 'Burger King, Dunkin' Donuts and community health care', *Health and Place* 6: 213–224.

Knox, P. (1995) *Urban Social Geography: An Introduction*, 3rd edition, Burnt Mill, Harlow: Longman Scientific and Technical.

Knox, P., Bohland, J. and Shumsky, N. L. (1983) 'The urban transition and the evolution of the medical care delivery system in America', *Social Science and Medicine* 17: 37–43.

Koenig, H. G. (1994) *Aging and God*, New York: Haworth Press.

Koenig, H. G. (1999) *The Healing Power of Faith: Science Explores Medicine's Last Great Frontier*, New York: Simon and Schuster.

Krause, E. A. (1977) 'The historical context of health', in *Power and Illness: The Political Sociology of Health and Medical Care*, New York: Elsevier.

Kuipers, J. C. (1989) ' "Medical discourse" in anthropological context: views of language and power', *Medical Anthropology Quarterly* 3: 99–123.

Kumar, S. V. (1983) *The Puranic Lore of Holy Water-Places*. New Delhi: Munshiram Manoharlal Publishers.

Laderman, C. (1987) 'The ambiguity of symbols in the structure of healing', *Social Science and Medicine* 24, 4: 293–301.

Lakoff, C. and Johnson, M. (1980) *Metaphors We Live By*, Chicago: University of Chicago Press.

Lane, S. and Meleis, A. (1991) 'Roles, work, health perceptions and health resources of women: a study in an Egyptian delta hamlet', *Social Science and Medicine* 33, 10: 1197–1208.

Lash, S. and Friedman, J. (eds) (1992) *Modernity and Identity*, Oxford: Basil Blackwell.

LaVeist, T. A. (1989) 'Linking residential segregation to the infant mortality race disparity in U.S. cities', *Sociology and Social Research* 73: 90–94.

LaVeist, T. A. (1990) 'Simulating the effects of poverty on the race disparity in post-neonatal mortality', *Journal of Public Health Policy* (Winter): 463–473.

Lawrence, C. (1995) *Medicine in the Making of Modern Britain 1700–1920*, London: Routledge.

Laws, G. and Radford, J. (1998) 'Place, identity, and disability: narratives of intellectually disabled people in Toronto', in R. A. Kearns and W. M. Gesler (eds) *Putting Health into Place: Landscape, Identity and Well-Being*, Syracuse, NY: Syracuse University Press.

Lazarus, E.S. (1988) 'Theoretical considerations for the study of the doctor–patient relationship: implications of a perinatal study', *Medical Anthropology Quarterly* 2, 1: 34–58.

Leatt, P. and Schneck, R. (1982) 'Work environments in different types of nursing
subunits', *Journal of Advanced Nursing* 7: 581–594.

Lee, P. R. (1982) 'Determinants of health', *Proceedings of the Conference of Health in the
'80s and '90s and Its Impact on Health Sciences Education*, Montebello, Quebec: Council
of Ontario Universities/Conseil des Universités de l'Ontario.

Leiderman, D. B. and Grisso, J. (1985) 'The gomer phenomenon', *Journal of Health and
Social Behavior* 26: 222–232.

Leiss, W., Kline, S. and Jhally, S. (1986) *Social Communication in Advertising: Persons,
Products and Images of Wellbeing*, Toronto: Methuen.

Lemert, C. (1997a) *Postmodernism Is Not What You Think*, Oxford: Blackwell.

Lemert, C. (1997b) *Social Things: An Introduction to the Sociological Life*, Lanham, MD:
Rowman and Littlefield.

Levin, J. S. (ed.) (1994) *Religion in Aging and Health: Theoretical Foundations and
Methodological Frontiers*, Thousand Oaks, CA: Sage.

Lewis, N. D. and Kieffer, E. (1994) 'The health of women: beyond maternal and child
health', in D. R. Phillips and Y. Verhasselt (eds) *Health and Development*, London:
Routledge.

Lewis, P. F. (1979) 'Axioms for reading the landscape', in D. W. Meinig (ed.) *The
Interpretation of Ordinary Landscapes*, New York: Oxford University Press.

Ley, D. (1981) 'Behavioral geography and the philosophies of meaning', in K. R. Cox
and R. C. Golledge (eds) *Behavioral Problems in Geography Revisited*, New York: Methuen.

Ley, D. (1988) 'Interpretive social research in the inner city', in J. Eyles (ed.) *Research
in Human Geography: Introductions and Investigations*, Oxford: Blackwell.

Ley, D. and Samuels, M. S. (1978) 'Introduction: contexts of modern humanism in
geography', in D. Ley and M. S. Samuels (eds) *Humanistic Geography*, Chicago: Maaroufa
Press.

Livingston, D. N. (1992) *The Geographical Tradition*, Oxford: Blackwell.

Longhurst, R. (1995) 'The geography closet in the body . . . the politics of pregnability',
Australian Geographical Studies 33: 214–223.

Longhurst, R. (1997) '(Dis)embodied geographies', *Progress in Human Geography* 21:
486–501.

Lorber, J. (1975) 'Good patients and bad patients: conformity and deviance in a general
hospital', *Journal of Health and Social Behavior* 16: 213–225.

McAuley, W. J., Pecchioni, L. and Grant, J. A. (2000) 'Personal accounts of the role
of God in health and illness among older rural African American and white residents',
Journal of Cross-Cultural Gerontology 15: 13–35.

McCormack, C. P. (1988) 'Health and the social power of women', *Social Science and
Medicine* 26: 677–683.

McGuire, M. B. (1983) 'Words of power: personal empowerment and healing', *Culture,
Medicine and Psychiatry* 7: 221–240.

Macintyre, S., Maciver, S. and Soomans, A. (1993) 'Area, class and health: should we
be focusing on places or people?', *Journal of Social Policy* 22, 2: 213–234.

Macintyre, S., Hunt, K. and Sweeting, H. (1996) 'Gender differences in health: are things really as simple as they seem?', *Social Science and Medicine* 42: 617–624.

McLafferty, S. (1986) 'The geographical restructuring of urban hospitals: spatial dimensions of corporate strategy', *Social Science and Medicine* 10: 1079–1086.

Malcolm, L. and Barnett, P. (1994) 'New Zealand's health providers in an emerging market', *Health Policy* 289: 85–100.

Mander, J. (1978) *Four Arguments for the Elimination of Television*, New York: Morrow Quill.

Maseide, P. (1991) 'Possibly abusive, often benign, and always necessary: on power and domination in medical practice', *Sociology of Health and Illness* 13, 4: 545–561.

Massey, D. (1997) 'A global sense of place', in T. J. Barnes and D. Gregory (eds) *A Global Sense of Place*, London: Arnold.

Massey, D. and Allen, J. (eds) (1984) *Geography Matters!*, Cambridge: Cambridge University Press.

Mayer, J. D. (1982) 'Relations between two traditions of medical geography: health systems planning and geographical epidemiology', *Progress in Human Geography* 6: 216–230.

Mayer, T. (1989) 'Consensus and invisibility: the representation of women in human geography textbooks', *Professional Geographer* 41, 4: 397–409.

Meade, M. S. and Earickson, R. J. (2000) *Medical Geography*, 2nd edition, New York: Guilford.

Meinig, D. W. (1979) 'The beholding eye: ten versions of the same scene', in D. W. Meinig (ed), *The Interpretation of Ordinary Landscapes*, New York: Oxford University Press.

Millard, A. V. (1994) 'A causal model of high rates of child mortality', *Social Science and Medicine* 38, 2: 253–268.

Mills, C. (1993) 'Myths and meanings of gentrification', in J. Duncan and D. Ley (eds) *Place/Culture/Representation*, London: Routledge.

Mills, W. J. (1982) 'Metaphorical vision: changes in Western attitudes to the environment', *Annals of the Association of American Geographers* 72: 237–253.

Miranda, P. N. (1989) *Terrorismo de estado: testimonio del horror en Chile y Argentina*, Colección Expediente Negro, Madrid: Editorial Sextante.

Mishler, E. G. (1984) *The Discourse of Medicine: Dialectics of Medical Interviews*, Norwood, NJ: Ablex.

Mitchell, D. (1995) 'There's no such thing as culture: towards a reconceptualization of the idea of culture in geography', *Transactions of the Institute of British Geographers* NS 20: 102–116.

Mitchell, D. (2000) *Cultural Geography: A Critical Introduction*, Oxford: Blackwell.

Mitchell, J. and Weatherly, D. (2000) 'Beyond church attendance: religiosity and mental health among rural older adults', *Journal of Cross-Cultural Gerontology* 15: 37–54.

Mohan, J. (1998) 'Explaining geographies of health care: a critique', *Health and Place* 4: 113–124.

Mohan, J. and Woods, K. J. (1985) 'Restructuring health care: the social geography of public and private health care under the British Conservative government', *International Journal of Health Services* 15: 197–215.

173

Moon, G. (1990) 'Conceptions of space and community in British health policy', *Social Science and Medicine* 30: 165–171.

Moon, G. and Brown, T. (1998) 'Space, place and health service reform', in R. A. Kearns and W. M. Gesler (eds) *Putting Health into Place: Landscape, Identity and Wellbeing*, Syracuse, NY: Syracuse University Press.

Morrill, R. L., Earickson, R. J. and Rees, P. (1970) 'Factors influencing distance traveled to hospitals', *Economic Geography* 46: 161–171.

Moss, P. (1997) 'Negotiating spaces in home environments: older women living with arthritis', *Social Science and Medicine* 45, 1: 23–33.

Munn, N. D. (1969) 'The effectiveness of symbols in Murngin rite and myth', in W. W. Spencer (ed.) *Forms of Symbolic Action*, Seattle: University of Washington Press.

Nast, H. (1994) 'Opening remarks: women in the field', *The Professional Geographer* 46: 54–66.

Nast, H. and Pile, S. (1998) 'Introduction: making places bodies', in H. Nast and S. Pile (èds) *Places through the Body*, London: Routledge.

Navarro, V. (1974) 'The underdevelopment of health or the health of underdevelopment: an analysis of the distribution of human health resources in Latin America', *International Journal of Health Services* 4: 5–27.

Neale, R. (1973) 'Society, belief, and the building of Bath, 1700–1793', in C. W. Chalkin and M. Havinden (eds) *Landscape and Society*, London: Longman.

Oman, D. and Reed, D. (1998) 'Religion and mortality among the community-dwelling elderly', *American Journal of Public Health* 88: 1469–1475.

Osborn, A. R. (1998) 'The regional distribution of alternative health care', in R. J. Gordon, B. C. Nienstedt and W. M. Gesler (eds) *Alternative Therapies: Expanding Options in Health Care*, New York: Springer.

Palka, E. J. (1995a) 'Coming to grips with the concept of landscape', *Landscape Journal* 14, 1: 63–73.

Palka, E. J. (1995b) 'Landscapes into place: an experiential view of Denali National Park', unpublished Ph.D. dissertation, Department of Geography, University of North Carolina.

Palka, E. J. (2000) *Valued Landscapes of the Far North: A Geographical Journey through Denali National Park*. Lanham, MD: Rowman and Littlefield.

Parker, R. (1983) *Miasma: Pollution and Purification in Early Greek Religion*, Oxford: Clarendon Press.

Parr, H. (1998) 'Mental health, ethnography and the body', *Area* 30: 28–37.

Parr, H. (1999) 'Mental health and the therapeutic geographies of the city: individual and collective negotiation', in A. Williams (ed.) *Therapeutic Landscapes: The Dynamic between Place and Wellness*, Lanham, MD: University of America Press.

Parr, H. (2000) 'Interpreting the "hidden social geographies" of medical health: ethnographies of inclusion and exclusion in semi-institutional spaces', *Health and Place* 6: 225–238.

Parr, H. (2001) 'New body-geographies: the embodied spaces of health and illness information on the Internet', *Environment and Planning D: Society and Space*.

Pavlovsky, E. N. (1966) *The Natural Nidality of Transmissible Disease*, Urbana: University of Illinois Press.

Peet, R. (1977) 'The development of radical geography in the United States', in R. Peet (ed.) *Radical Geography: Alternative Viewpoints on Contemporary Social Issues*, London: Methuen.

Philo, C. (1987) 'Fit locations for an asylum: the historical geography of the nineteenth century "mad-business" in England as viewed through the pages of the *Asylum journal*', *Journal of Historical Geography* 13: 398–415.

Philo, C. (2000) 'Post-asylum geographies: an introduction', *Health and Place* 6, 3: 135–136.

Pile, S. and Thrift, N. (1995) 'Mapping the subject', in S. Pile and N. Thrift (eds) *Mapping the Subject: Geographies of Cultural Transformation*, London: Routledge.

Pinfold, V. (2000) 'Building up safe havens . . . all around the world: users' experiences of living in the community with mental health problems', *Health and Place* 6: 201–212.

Pocock, D. C. D. (1981) 'Introduction: imaginative literature and the geographer', in D. C. D. Pocock (ed.) *Humanistic Geography and Literature: Essays on the Experience of Place*, London: Croom Helm.

Poland, M. L. (1984) *Unemployment, Stress, and Infant Mortality: Detroit*, Chapel Hill: Institute of Nutrition, University of North Carolina.

Porteous, D. J. (1985) 'Smellscape', *Progress in Human Geography* 9: 356–378.

Pratt, G. (2000a) 'Post-structuralism', in R. J. Johnston, D. Gregory, G. Pratt and M. Watts (eds) *The Dictionary of Human Geography*, 4th edition, Oxford: Blackwell.

Pratt, G. (2000b) 'Body, geography and', in R. J. Johnston, D. Gregory, G. Pratt and M. Watts (eds) *The Dictionary of Human Geography*, 4th edition, Oxford: Blackwell.

Pratt, G. (2000c) 'Geography and sexuality', in R. J. Johnston, D. Gregory, G. Pratt and M. Watts (eds) *The Dictionary of Human Geography*, 4th edition, Oxford: Blackwell.

Pred, A. (1984) 'Place as historically contingent process: structuration and the time-geography of becoming places', *Annals of the Association of American Geographers* 74: 279–297.

Pred, A. (1989) 'The locally spoken word and local struggles', *Environment and Planning D: Society and Space* 7: 211–233.

Price, L. (1987) 'Ecuadorian illness stories: cultural knowledge in natural discourse', in D. Holland and N. Quinn (eds) *Cultural Models in Language and Thought*, Cambridge: Cambridge University Press.

Price, M. and Lewis, M. (1993) 'The reinvention of cultural geography', *Annals of the Association of American Geographers* 83, 1: 1–17.

Relph, E. (1976) *Place and Placelessness*, London: Pion.

Robinson, R. and LeGrand, J. (eds) (1993) *Evaluating the NHS Reforms*, London: King's Fund Institute.

Rodaway, P. (1994) *Sensuous Geographies: Body, Sense, Place*, London: Routledge.

Rogers, E. M. (1979) 'Network analysis of the diffusion of innovations', in P. W. Holland and S. Leinhardt (eds) *Perspectives on Social Network Research*, New York: Academic Press.

Rose, G. (1993) *Feminism and Geography: The Limits of Geographical Knowledge*, Minneapolis: University of Minnesota Press.

Rowles, G. D. (1978) 'Reflections on experiential field work', in D. Ley and M. S. Samuels (eds) *Humanistic Geography: Prospects and Problems*, London: Croom Helm.

Rowntree, L. B. (1988) 'Orthodoxy and new directions: cultural/humanistic geography', *Progress in Human Geography* 12, 4: 575–586.

Rowntree, L. B., Foote, K. E. and Domosh, M. (1989) 'Cultural geography', in G. L. Gaile and C. J. Willmott (eds) *Geography in America*, Columbus. OH: Merrill.

Rubel, A. J. and Hass, M. R. (1995) 'Ethnomedicine', in C. F. Sargent and T. M. Johnson (eds) *Medical Anthropology: Contemporary Theory and Method*, Westport, CT: Praeger.

Rundall, T. G. and McClain, J. O. (1982) 'Environmental selection and physician supply', *American Journal of Sociology* 87, 5: 1090–1112.

Sack, R. D. (1976) 'Magic and space', *Annals of the Association of American Geographers* 66: 309–322.

Sack, R. D. (1992) *Place, Modernity and the Consumer's World*, Baltimore: Johns Hopkins Press.

Said, E. (1978) *Orientalism: Western Conceptions of the Orient*, New York: Pantheon.

Sanjek, R. (1990) *Fieldnotes: The Makings of Anthropology*, Ithaca, NY: Cornell University Press.

Sayer, A. (1992) *Method in Social Science: A Realist Approach*, 2nd edition. New York: Routledge.

Scarpaci, J. L. (ed.) (1989a) *Health Services Privatization in Industrial Societies*, New Brunswick: Rutgers University Press.

Scarpaci, J. L. (1989b) 'Introduction: the theory and practice of health services privatization', in J. L. Scarpaci (ed.) *Health Services Privatization in Industrial Societies*, New Brunswick, NJ: Rutgers University Press.

Scarpaci, J. L. (1999) 'Healing landscapes: revolution and health care in post-socialist Havana', in A. Williams (ed.) *Therapeutic Landscapes: The Dynamic between Place and Wellness*, Lanham, MD: University Press of America.

Schneider, D. (1998) 'Demand for alternative therapies: the case of childbirth', in R. J. Gordon, B. C. Nienstedt and W. M. Gesler (eds) *Alternative Therapies: Expanding Options in Health Care*, New York: Springer.

Secretary of State for Health (1989) *Working for Patients*, HMSO: London.

Seligman, M. E. P. (1970) 'On the generality of the laws of learning', *Psychological Review* 77: 406–418.

Shields, R. (1992) *Lifestyle Shopping: The Subject of Consumption*, London: Routledge.

Shorter, J. (1985) 'Accounting for place and space', *Environment and Planning D: Society and Space* 3: 447–460.

Sibley, D. (1995) *Geographies of Exclusion*, London: Routledge.

Singh, J. (1990) 'A multifacet typology of patient satisfaction with a hospital stay', *Journal of Health Care Marketing* 19, 4: 8–21.

Skelton, T. and Valentine, G. (1998) *Cool Places: Geographies of Youth Cultures*, London: Routledge.

Smart, B. (1996) 'Postmodern social theory', in B. S. Turner (ed.) *The Blackwell Companion to Social Theory*, Oxford: Blackwell.

Smith, C. and Giggs, J. (eds.) (1988) *Location and Stigma: Contemporary Perspectives on Mental Health and Mental Health Care*, London: Unwin Hyman.

Smith, N. and Katz, C. (1993) 'Grounding metaphor: towards a spatialized politics', in M. Keith and S. Pile (eds) *Place and the Politics of Identity*, London: Routledge.

Smith, S. J. (1994) 'Soundscape', *Area* 26: 232–240.

Soja, E. (1986) 'Taking Los Angeles apart: some fragments of a critical human geography', *Environment and Planning D: Society and Space* 4: 255–272.

Sontag, S. (1978) *Illness as Metaphor*, New York: Vintage Books.

Spencer, C. and Blades, M. (1986) 'Pattern and process: an essay on the relationship between behavioral geography and environmental psychology', *Progress in Human Geography* 10: 230–248.

Spradley, J. P. (1980) *Participant Observation*, New York: Holt, Rinehart and Wilson.

Staiano, K. V. (1979) 'A semiotic definition of illness', *Semiotica* 28: 107–125.

Staiano, K. V. (1981) 'Alternative therapeutic systems in Belize: a semiotic framework', *Social Science and Medicine* 15B: 317–332.

Stephenson, P. H., Elliott, S. J., Foster, L. T. and Harris, J. (1995) *A Persistent Spirit: Towards Understanding Aboriginal Health in British Columbia*, Canadian Western Geographical Series vol. 31: Department of Geography, University of Victoria.

Stern, J. P. (1995) 'Relativity in and around *The Magic Mountain*', in M. Minden (ed.) *Thomas Mann*, London: Longman.

Stock, R. (1986) ' "Disease and development" or "the underdevelopment of health": a critical review of geographical perspectives on African health problems', *Social Science and Medicine* 23: 789–700.

Suchman, A. L. and Matthews, D. A. (1988) 'What makes the patient–doctor relationship therapeutic? Exploring the connexional dimension of medical care', *Annals of Internal Medicine* 108: 125–130.

Tambiah, S. J. (1968) 'The magical power of words', *Man* 3, 2: 175–208.

Techatraisak, B. and Gesler, W. M. (1989) 'Traditional medical practitioners in Bangkok, Thailand', *Geographical Review* 79, 2: 172–182.

Thurber, C. and Malinowski, J. (1999) 'Summer camp as a therapeutic landscape', in A. Williams (ed.) *Therapeutic Landscapes: The Dynamic between Health and Wellness*, Lanham, MD: University Press of America.

Todd, A. D. and Fisher, S. (1993) *The Social Organization of Doctor–Patient Communication*, 2nd edition, Norwood, NJ: Ablex.

Townsend, P., Simpson, D. and Tibbs, N. (1985) 'Inequalities in health in the city of Bristol: a preliminary review of statistical evidence', *International Journal of Health Services* 15, 4: 637–663.

Travers, M. (1992) *Thomas Mann*, New York: St Martin's Press.

Tuan, Y.-F. (1976) 'Humanistic geography', *Annals of the Association of American Geographers* 66: 266–276.

Tuan, Y.-F. (1979) *Landscapes of Fear*, New York: Pantheon Books.

Tuan, Y. -F. (1984) *Dominance and Affection: The Making of Pets*, New Haven, CT: Yale University Press.

Tuan, Y.-F. (1988) 'The city as a moral universe', *Geographical Review* 78: 316–324.

Tuan, Y.-F. (1991) 'Language and the making of place: a narrative-descriptive approach', *Annals of the Association of American Geographers* 81, 4: 684–696.

Tuckett, D., Boulton, M., Olson, C. and Williams, A. (1985) *Meetings between Experts: An Approach to Sharing Ideas in Medical Consultations*, London: Tavistock.

Turner, V. W. (1975) 'Symbolic studies', *Annual Review of Anthropology* 4: 145–161.

Turow, J. (1997) 'James Dean in a surgical gown: making TV's medical formula', in L. Spiegel and M. Curtin (eds) *The Revolution Wasn't Televised: Sixties Television and Social Conflict*, New York: Routledge.

Turshen, M. (1977) 'The political ecology of disease', *Review of Radical Political Economy* 9: 45–60.

Ulrich, R. S. (1983) 'View through window may influence recovery from surgery', *Science* 224: 420–421.

Vlach, J. M. (1993) *Back of the Big House: the Architecture of Plantation Slavery*, Chapel Hill, NC: University of North Carolina Press.

Wagner, P. (1975) 'The themes of cultural geography rethought', *Yearbook, Association of Pacific Coast Geographers* 37: 7–14.

Wagner, P. and Mikesell, M. (1962) 'The themes of cultural geography', in P. Wagner and M. Mikesell (eds) *Readings in Cultural Geography*, Chicago: University of Chicago Press.

Waitzkin, H. (1989) 'A critical theory of medical discourse: ideology, social control, and the processing of social context in medical encounters', *Journal of Health and Social Behavior* 30: 220–239.

Warf, B. (1994) 'Review of place, modernity and the consumer's world (R. D. Sack)', *Geographical Review* 84: 107–109.

Weidman, H.-H. (1977) 'On Mitchell's changing others', *Medical Anthropology Newsletter* 8, 14: 25.

Weigand, H. J. (1965) '*The Magic Mountain*': A Study of Thomas Mann's Novel 'Der Zauberberg', Chapel Hill: University of North Carolina Press.

Weil, C. and Weil, J. (1988) 'Differential acceptance of oral rehydration therapy in a multi-ethnic setting in Bolivia: a research strategy and preliminary findings', Paper presented at the Geography Workshop, University of Chicago, May 1988, used by permission of the authors.

Wennberg, J. and Gittelsohn, A. (1982) 'Variations in medical care among small areas', *Scientific American* 245: 120–133.

Wennemo, I. (1993) 'Infant mortality, public policy and inequality – a comparison of 18 industrialised countries 1950–85', *Sociology of Health and Illness* 15, 4: 429–446.

WHO (1946) *Constitution*, New York: World Health Organization.

Widdowfield, R. (2000) 'The place of emotions in academic research', *Area* 32: 199–208.

Wilkinson, R. (1996) *Unhealthy Societies*, London: Routledge.

Wilkinson, R. G. (1990) 'Income distribution and mortality: a "natural" experiment', *Sociology of Health and Illness* 12, 4: 391–412.

Williams, A. (ed.) (1999) *Therapeutic Landscapes: The Dynamic between Wellness and Place*, Lanham, MD: University of Press of America.

Williams, M. A. (1988) 'The physical environment and patient care', *Annual Review of Nursing Research* 6: 61–84.

Williams, R. (1993) 'Culture is ordinary', in A. Gray and J. McGuigan (eds) *Studying Culture: An Introductory Reader*, London: Edward Arnold.

Williamson, J. (1978) *Decoding Advertisements: Ideology and Meaning in Advertising*, London: Marion Boyars.

Wilton, R. D. (1996) 'Diminished worlds? The geography of everyday life with HIV/AIDS', *Health and Place* 2, 2: 69–83.

Wilton, R. D. (1999) 'Qualitative health research: negotiating life with HIV/AIDS', *The Professional Geographer* 51, 2: 254–264.

Wolch, J. and Philo, C. (2000) 'From distributions of deviance to definitions of difference: past and future mental health geographies', *Health and Place* 6, 3: 137–157.

Zola, I. K. (1972) 'Medicine as an institution of social control', *Sociological Review* 20: 487–504.

INDEX

Printed in the United States
by Baker & Taylor Publisher Services